D0894334

IEE Management of Technology Series 6

Series Editor: G. A. Montgomerie

MANAGEMENT FOR ENGINEERS

Revised second edition

Other volumes in this series

Volume 1	Technologies and markets J. J. Verschuur
Volume 2	The business of electronic product development Fabian Monds
Volume 3	The marketing of technology C. G. Ryan
Volume 4	Marketing for engineers J. S. Bayliss
Volume 5	Microcomputers and marketing decisions L. A. Williams
Volume 6	Management for engineers D. L. Johnston
Volume 7	Perspectives in project management R. N. G. Burbridge (editor)
Volume 8	Business for engineers B. C. Twiss
Volume 9	Exploitable U.K. research for manufacturing industry G. Bryan (editor)

MANAGEMENT FOR ENGINEERS

Revised second edition

DL Johnston

Peter Peregrinus Ltd on behalf of the Institution of Electrical Engineers

Published by: Peter Peregrinus Ltd., London, United Kingdom

Revised Second Edition 1989

British Library Cataloguing in Publication Data

Johnston, Denis L.
Management for engineers.—Rev. 2nd Ed.
(IEE management of technology series; v. 6).
1. Engineering—Management
I. Title II. Series
658′.002462 TA190

ISBN 0 86341 161 4

Printed in England by Short Run Press Ltd., Exeter

Contents

Foreword

Management expertise and technological skills do not always sit happily together in industry, and anything that brings these disciplines together is to be welcomed. Therein lies the value of Mr. Johnston's book.

Scientists, engineers and managers in industry are, or should be, working for common goals . . . to meet the needs of consumers and to raise the level of well being in society. That is the justification for their training, and indeed for their existence.

Management is less about bossing people around than about agreeing objectives and achieving them cost effectively. Above all, it requires leadership to get the best out of oneself as manager and whatever human, material and financial resources for which the manager is responsible.

In this respect, engineering is no different from any other function in business. But engineers frequently need encouragement to recognise a wider responsibility in the management of resources, and then need to be helped to develop the personal skills which will enable them to handle that responsibility.

"Management for Engineers" will stimulate the recognition by professional engineers of the importance of people, of marketing, and of financial control. And it will encourage independent thought, and thus the opportunity for the acquisition of a wider perception of the rôle of engineering in the modern world.

Lord Weinstock

Introduction

Surveys have shown that the majority of professional engineers or scientists are required to assume a degree of management responsibility as they approach mid-career, and, for some, the opportunity may come earlier.

This should present no problem if one prepares in anticipation: it makes no contribution to a career to decline, or to fail when added responsibilities are thrust upon us.

An introduction is provided to the wider framework in which a team leader, a departmental or a general manager operates. This role is effectively to integrate and lead the work of a specialist team, ranging from those in a small project up to the whole staff of a major enterprise. The professional engineer, in the earlier part of a career, faces the problem that, as a specialist, he or she has little appreciation of the contributions of colleagues in disciplines other than technology. There is a need to grow towards becoming a generalist. This is a suitable career progression, bearing in mind the inevitable rate of obsolescence in one's own earlier technical or scientific specialism.

In running an established business, starting up a new venture, or managing a project, just one blind spot in the awareness of the leader can bring disaster. The increasing pace of change in the technological industries makes it impracticable in a single career to embrace personally the whole spectrum of management experience; so some short cuts are essential. This dynamic situation requires an awareness of what has gone before, in order to extrapolate and anticipate what may come.

The presentation in this book will ensure awareness of the principal management disciplines and techniques, together with guidelines so that the reader can seek more detailed information whenever the need arises. Such sources include other books in the Management of Technology series, the work done by the IEE's Management & Design Division since its formation in 1980, and some quotations from distinguished professional engineers who have been successful in making the transition to higher management.

In this book and its bibliographies will be found the main elements required when preparing for the Diploma in Engineering Management introduced by the

I.Mech.E. and supported by the IEE, in particular for the content of Part 1 of the Syllabus. Candidates for Part 2 are also provided for in the volume of this Series, 'Business for Engineers' by B.C. Twiss. Corporate Members of the Engineering Institutions, who have been awarded the Diploma in Engineering Management, are eligible for the statutory (full) membership of the British Institute of Management.

The literature of management is so extensive today that only a specialist in the subject has time to explore and assess the material. The conceptual approach used in this book is unusual in this field, but will be familiar to professional engineers in their technological and scientific work. They are well able to follow this short cut to the appreciation of the parameters of the business disciplines.

A short representative Reading List will be found at the end of each main Section. Specialised references in the form 'DRUCKER (1954)' will be found listed alphabetically in corresponding subsections of the main Bibliography (Section 7). These can provide 'signposts' for the pursuit of particular subjects in greater depth.

Chapter 1

Organisation

1.1 Summary

The reader probably started his career as a technical specialist, and is now ready for a wider appreciation of:

- what his colleagues do, in other functions than his own
- a view of his company or organisation as a whole
- its place in the business community, at home and internationally

An engineer or scientist, moving towards management responsibilities as mid-career is approached, is not abandoning the first profession, but forming links with other colleagues who make up the complete management team. From the new position of responsibility it will be possible to establish conditions in which better technical and scientific work can be done.

The manager in engineering has a more challenging job than managers in some other fields, primarily because of the complexity of the processes, and the continuous condition of innovation and change. To offset the exacting nature of the task of management and direction, there is the advantage of possessing a high degree of numeracy, and the habit of and inclination to achieve results.

In this first Chapter of the book, a brief historical perspective is sketched, not only for its inherent interest, but because it demonstrates that continuing change is the norm. We have two seeming contrary trends: integration and amalgamation of businesses, on the one hand, and on the other hand, a move to smaller operational units, and conditions favourable to various forms of individual enterprise and new ventures.

After a brief review of the development of what are now conventional organisational hierarchies, the impact of integrated networks and automated manufacture is highlighted, coupled with the rather different ideas of the role of authority that we have gleaned from the successes of the Japanese.

The importance is stressed of formulating clear strategies and objectives, and using them in a dynamic way as reference points, so that the process of management works as a system, stabilised and monitored by feedback loops.

An assessment is made of the variants that may be met of 'management by objectives' (MbO), and the contribution to its practice of Peter Drucker, who wrote some 30 years ago, that '. . . in management there is a tremendous gap between the knowledge and performance of the leaders, and of the rest. . .'. In a changing scene this is still so today.

In providing an overview of 'organisation', an attempt is made to increase the reader's awareness of the whole field. What is written here will begin to fill any gaps or blind spots, and a fairly full bibliography (Section 7.1) provides a lead to further sources of information. The references are not exhaustive, but in general are recent and can point the reader to earlier work. The roles of the several business functions are outlined, and most of them now have their own professional institutes, with an organised 'body of knowledge'.

The special skill of the engineering manager is in being accustomed to manage change, and the project concept is seen as having a wider meaning for the future, both when applied on a modest scale, and when coupled with the emerging ideas of systems management, based on the experience of the very large engineering projects.

The Chapter concludes with reference to the need for awareness and sensitivity to the cultural dimension, both within one's own organisation and country, but also when exercising the opportunities that arise to work internationally.

1.2 Historical note

1.2.1 Innovators in management thinking

It is useful to step back, in order to see the total historic perspective in which the practice of 'management' has grown up. It has been said:

> study the past, if you would divine the future (*Confucius*, 551–479 BC)

No one can be certain of the future in all its detail, but history can persuade us of the dynamic nature of social and economic development, particularly where technology intervenes.

As in scientific innovation and technical development, ideas have appeared and were known to a limited circle, well before they became more generally available.

A few of the major early contributions to the body of management knowledge are mentioned below, together with some dates, based on the assessment by Urwick (1956):*

Henry R Towne, a manufacturer in USA, presented in 1886 a pioneer 'management' paper to the American Society of Mechanical Engineers (ASME) on 'The engineer as an economist'.

---

*Urwick (1956) writes of 70 of the early pioneers of management, and the story is brought up to date by Pugh's (1984) selection of 'readings' on management, in a Penguin edition.

Slater Lewis, whose book 'The commercial organisation of factories' (1896) introduced the graphic concept of the organisation tree.

A M Church's 1901 paper, which was the starting point for modern cost and management accounting.

F W Taylor and H L Gantt had collaborated in the American engineering industry for a number of years before they published in 1901–03 their method of work measurement (then known as 'scientific management' because it involved measurement, not opinion), and its use by Gantt to programme and schedule in manufacture, and in task and bonus schemes.

The Gilbreths, Frank and Lillian, were active contributors from the turn of the century, and she lived until 1973. He specialised in improving methods in construction and engineering, extending the work of Taylor and Gantt. She applied her training in psychology to a better understanding of the motivation of workpeople.

Anne Shaw, a British student of Lillian Gilbreth, did excellent work with female staff at Metropolitan Vickers' factories during the Second World War, and afterwards applied the experience widely in industry. By then, the 'Hawthorne experiments' by Mayo at the factories of Western Electric were well known.

Henri Fayol's classic work on administrative systems was begun well before the turn of the century, first presented as a lecture in 1900 and later printed in France (1908). The first edition in English was printed in 1929 (Fayol, 1929).

B Seebohm Rowntree was in close touch and well aware of pioneer work by others internationally. From the early 1920s onwards the Rowntree Cocoa Works at York was systematically reorganised according to the new approach to management that he had gleaned from every available source (Briggs, 1961).

Oliver Sheldon had his whole career at Rowntree's firm, and wrote a definitive book on the concepts of management (1923), at a time when the British 'management literature' was slight by comparison with that of the USA. He was well aware of the more advanced thinking, and related it to the humanitarian approach of Rowntree's company, where he worked and rose to become Chairman.

Mary Parker Follett had a major impact in England, when she lectured in the late 1920's on the inter-personal relations in industrial organisation, and on the analysis of conflict and authority. She was a major influence on the American Management Association (AMA). She said in 1925:

> The next step business management should take is to organise the body of knowledge on which it should rest — Urwick (1956)

1.2.2 Arrival of professionalism

The years of the First World War had shown the need for professionalism in the conduct of industrial and business functions, and were followed by the starting up of:

Federation of British Industries (1918)
Industrial Welfare Society (1918), now The Industrial Society

Institute of Chartered and Works Accountants (1919)
Institute of Industrial Administration (1919), merged with BIM
National Institute of Industrial Psychology (1920)
Institution of Production Engineers (1921)

It was the view of Urwick who knew him well, that Rowntree had 'a greater influence than any other businessman... towards guiding the country to a... more enlightened view of... business leadership'.

In some areas ideas spread slowly. While Taylor gave his first lecture to ASME in 1895, and in a published Paper in 1903, and a book in 1910, a comparable paper to the I.Mech.E. in Britain did not appear until that by Fontaine, Walker and Spencer (1958), and to the IEE by Mayer (1973). But both Institutions did make factory organisation an option in their examination syllabus, the I.Mech.E. as early as 1913.

Taylor's terminology of 'scientific management' became bad news, and is never used today: what he meant was that targets, actions and decisions were based on quantified measurement, rather than on custom, opinion or wish fulfilment. Time standards came to be associated with negotiation of pay formulas, losing sight of two principles that Taylor had stated (page 74 in his book):

- both sides (management and men) must take their eyes off the division of the surplus... and turn their attention towards increasing the size of the surplus
- both sides must recognise as essential the substitution of exact scientific investigation and knowledge for the old individual judgement or opinion.

He summed up, that all concerned must have a common purpose and a common method.

In summary, Taylor developed 'management' from the shop floor upwards, while Fayol worked from the top downwards. The I.Mech.E. (founded in 1847) made a contribution at the manufacturing level, but its *Proceedings* contains only the one paper cited in the decade of the 1950s, and one other in the 1960s. From 1970 the subject received frequent attention. At the IEE a Professional Group on Engineering Management was established in 1970, just a century after the Institution was founded, and the Management & Design Division was set up in 1980.

The tradition at the Institution of Civil Engineers has always been to manage construction work, and practical experience has long been a requirement for corporate membership. The I.Prod.E.'s interests have moved upwards towards general management, and they pioneered with Conferences on Automation in 1955 and 1957 (I.Prod.E., 1955, 1957).

1.2.3 The Business Schools

The Business Schools made an early but slow start. In 1881 was founded the Wharton School of Finance and Commerce at the University of Philadelphia. By 1901 there were eight Business Schools in the USA, 20 by 1910, over 100 by 1924. By 1980 over 500 were turning out more than 50 000 qualified students a year, as noted in the survey published by the IEE (1982), on Business Schools.

In the UK the Balfour Committee of 1924 drew attention to these developments, a theme taken up again in the Urwick report of 1946, and by Franks in 1963. This led to the formation of the first two British Business Schools in 1965 (Manchester and London), but it should not be overlooked that some of the needs had been met earlier by The Regent Street Polytechnic, some regional Schools of Commerce (notably Manchester), the University commerce courses, and by the founding in 1945 of the Henley Staff College.

The Institute of Directors had been established in 1903, in the wake of the 1900 Companies Act and other legislation. By 1910 it had 1358 members, and it currently has 33 000.

On the recommendations of the Baillieu Committee of 1946, the British Institute of Management began operations in 1948. The American Management Association (AMA) had been founded in 1923 out of a merger of several specialised and regional associations. By 1929 it operated fully nationally, but did not recover its full momentum until after the economic depression of the early 1930's.

1.2.4 Social conscience

The lack of a general social conscience in industry and business in the 19th century is probably sketched quite accurately in Dickens's stories of London, and in Upton Sinclair's books, such as 'The Jungle' (1906), placed in Chicago.

Rowntree did much to influence changes in attitude in the UK, and collected round him a circle of like-minded people: he had a position of influence during the First World War, when he was appointed Director of the Industrial Welfare Department of the Ministry of Munitions, on the strength of his work in the family firm. Educated at the Friends' School, York, he became interested in social research, and was able to persuade his fellow Directors to develop a social philosophy and policy. As early as 1906 there was a pension scheme, with retirement at 65 for men and at 60 for women.

There were other isolated examples of paternalistic or enlightened industrialists: Cadbury's Bournville model town, Lever's Port Sunlight, and Howard's 'Garden Cities' of Letchworth and Welwyn. But most notable as a pioneer was Sir Titus Salt of Bradford, who, in 1853, opened a model community of weaving mills, housing and social amenities at Saltaire, recently restored and now open to the public.

John Ruskin recorded that the objective was to provide constant employment for a thousand workers who never drank, never went on strike and always attended church on Sundays. Salt showed real concern for his employees' welfare, but held their wages in line with those of competitive mills.

1.2.5 Slow to change

Despite these examples of advanced thinking well ahead of their time, H G Wells could write in his book 'The work, wealth and happiness of mankind' (chapter 11):

> In England we have come to rely upon a comfortable time-lag of fifty years

> or a century intervening between the perception that something ought to be done and a serious attempt to do it.

An assessment by Dr Keeble (1984) of London School of Economics is that British owner-managers were slow to accept the idea of professional managers, men with expertise but little or no share-holding in a company. Both management education and the business graduate were in little demand in the UK during the first half of this century. In the late 19th century, when family firms predominated, a son or a nephew could be expected to take over, and very few firms had the mechanism for propelling the most able employees to the top: the assumption was that all employees should start at the bottom and 'come up the hard way'.

In many cases the firm was an extension of the family: the firm's hierarchy replicated the family and the head of the family was, by tradition, the head of the firm. Personal recommendation seemed much the best way of discovering the man best suited to a family environment. Dr Keeble quotes James Bowie's comment in 1930:

> Nepotism, by denying its reward, is perhaps the biggest single obstacle in Britain, against the improvement of industrial practice.

Writing in 1982, Christopher Lorenz identified a general 'anti-industrial culture' as lying at the roots of the British malaise and poor economic performance over the last 100 years, and quoted several recent books, in particular the research of the Cambridge historian, Correlli Barnett. In 'The Audit of War', Barnett (1986) takes the industrial mobilisation for the Second World War as his source of the evidence that, while there was brilliance at the level of research and design, the newer industries had hardly emerged from the cottage stage. Basing his study on recently de-classified government wartime files, a general picture of inadequate management and poor Union co-operation is revealed.

When the USA began to contribute to the Allied cause after Pearl Harbour, they built very quickly vast production facilities, such as Kaiser's shipyards and the Willow Run aircraft plant. This was done with a manufacturing professionalism that appeared to be absent in the UK. There was the opportunity to catch up when, as is recorded by Tisdall (1982):

> Sir Stafford Cripps, initially as President of the Board of Trade, created a framework which set the pace for the expansion of management and management services including consulting, which lasted for the next 25 years.

It may be noted that, for a politician, Cripps was highly interdisciplinary, being both a scientist and a lawyer: he had taken the major role in the definitive patent case over the 'pentode' valve in the 1930's. He set up the Angle-American Council on Productivity which, during the period 1948–52, sent to the USA a large number of specialist teams, comprising both management and trade unionists. This was a unique opportunity to absorb the 'state of the art' in many fields, and some of the lessons were learnt and implemented, but some were not. In the 1980s the Japanese are equally willing to show what they are doing, but, over the cups of *sake*, have been known to say that they do not really expect that the occidentals are able to catch up.

1.2.6 Strengths and weaknesses in research, design and development

The effective management of innovation is a recurrent theme. The UK was early in the field with GEC's Industrial Research Laboratories, opened up by the late Sir Clifford Paterson in 1919 at East Lane, Wembley, which set the organisational style for many other establishments of that era. The Bell Telephone Laboratories, originally in New York City, were founded in 1925, and among many contributions is credited with the origination of 'systems engineering', which is now integrated into the practice of management of complex projects: (M'Pherson, 1986).

The pattern of industrial research has changed since the British government stimulated the activity during the First World War by setting up the Department of Scientific & Industrial Research (DSIR) to fill the gaps in the skills then available in the country, such as the manufacture of dyestuffs and optical instruments, for which Germany had been a source.

By the early 1920s there were 20 research associations for different industries, funded 50/50 by their industry and by government. The number reached its peak of 48 in 1970, and has now declined.

In their place today we see the growth of 'clubs', a free interchange of information within consortia on basic technology, usually as a programme on a project basis, of defined duration. The 'clubs' are administered by an appropriate body: for example, Harwell is the base for a number, and the Department of Trade & Industry looks after the Alvey Directorate (the programme on the fifth generation computer). In Japan, MITI pioneered this role as described by Johnson (1982), and in the EEC there are programmes such as ESPRIT and Eureka for industrial collaboration in high technology, involving 18 countries.

1.3 Growth of the large Corporations

1.3.1 Corporations and people

The large Corporations of the earlier part of this century developed from the tendency to work towards monopolies in each industry, and because of the economies of scale that applied to the less flexible technologies of the era. Their influence increased in the mid-century as they developed into multinationals.

Although international trade stretches back into history, there was a buccaneering flavour in the days of Elizabeth I. The theme of the roles, rights and responsibilities of the international corporation' was first seriously debated at the 1969 Congress of the International Chambers of Commerce, held in Istanbul, and Tugendhat (1971) has developed this theme, and provides a number of case histories.

The impact on people of the large Corporations has been generally favourable, as they offer better than average conditions of employment, but not as high a security of tenure in employment as people have come to expect in the public service.

For the younger person the large firms provide good opportunities for training and experience, either informally by observation of the good and bad features of the establishment, or through the formal training and management development opportunities that a smaller company rarely can provide.

The best training, and often the way to the top of the Corporation, is through a progression of internal job changes, both laterally to gain experience of other functions, and by upwards promotion to broader responsibilities. The multinationals provide chances of international experience or, as it is called 'management transfer', (the subject of a special issue of *International Management* in May 1986). Bray (1974) made a study of the careers and job attitudes of 270 graduates recruited by the Bell Telephone System, which provides a good model for making a similar study in any large organisation, and it highlighted the importance of the initial selection process.

1.3.2 Transfer of Experience

The mature countries have provided a degree of stability that inhibits personal job mobility, thereby foregoing one of the most effective ways to transfer experience. At a colloquium on 'Systems Engineering' M'Pherson (1986), it was said that it takes too long to gain experience in managing major projects; a person may be involved in only three in 30 years. Ways must be found to teach the skills to younger people over 3–5 years.

The multinationals perform a valuable function in demonstrating an advanced way of doing business in the newly industrialised countries (NIC), and in the Third World, for example by ITT, Xerox and IBM. Their style of operation may be less suitable for locally based businesses but, as a prestige model, has a stimulating effect.

The formal approach to transfer of experience comes when a particular function becomes professionalised, and a recognised 'body of knowledge' is systematised and taught. There is a negative aspect to this, as practices may become 'frozen', and innovation inhibited; to some extent the accountancy profession has had this problem during the period when computer methodology has been forcing the pace of change.

In earlier history it was usual to impose the parent country's culture on overseas branches of a business. Very real efforts are made nowadays to make use of the best features of the local culture, and to minimise the proportion of expatriates in senior positions. An excellent account of the experience of the Honeywell company was given by George (1983): there are 50 subsidiaries, each headed by a national, and the group applies worldwide ethical standards. There is greater wisdom today in the conduct of overseas business, by comparison with the earlier time of, for example, ITT, as told by Sampson (1973).

A great deal has been written about the Japanese, and Livingstone (1984) tells of a year's exchange experience, arranged between British Telecom and a company in Japan. When Japanese companies operate abroad they adjust their style to the country, but do not relax their very high standards.

In working up to their now advanced capability the Japanese made excellent use of information and advice equally available to Western businesses. Their outstanding success has been to demonstrate high levels of quality without excess costs, and with high productivity. This was largely based on taking seriously the advice of Western consultants such as Dr. W. Edwards Deming, who, at the time of writing, is still active at 85, and J.M. Juran who was originally with the Western Electric Company, and who published his work freely; for example, see Juran (1935). If some countries seem to have advanced very quickly it is because excellent people have been trained, very often spread over the best of the world's universities, and they have then applied themselves to identify and select the most appropriate technologies, from wherever.

Within a mature country we tend to feed on our own resources, but an emerging country can seek more widely in the whole world market. This applies both to the purchase of equipment, and to the acquiring of 'know-how' licences. In some fields the costs of development and tooling-up are such that the principal international competitors have to make co-operative arrangements, as in some types of aero engines and with semiconductors, and where common technical standards are essential for compatibility of equipment or systems. To succeed today and in the future, products and systems must be in 'world class'.

1.3.3 Trends in integration and mergers

The wave of business mergers in recent times is the consequence of greater concern about the cost of money, and hence the optimum uses of resources, the generation of cash flow, and the disposal of activities that do not fulfil certain criteria. The mergers that fail to achieve results tend to be those where the people in command are entering a new field for which they have no 'gut feel': on the other hand, there have been many old established businesses where a shake-up was overdue, and historic but unprofitable activities needed to be pruned.

Increasingly a business group is happy to delegate responsibility for the operation of profit centres to those on the spot; so the engineer-manager is reasonable isolated from corporate changes, provided that his unit is successful in generating a steady cash flow. The process of pushing responsibility downwards has not been easy in some of the larger and previously monolithic corporations. In companies such as ICI and ICL it has amounted to a major cultural upheaval, which has had to be stimulated to bring about the necessary changes in attitudes. Each large company has its own internal culture, and effective integration after a merger can be difficult.

The scale of recent integration can be appreciated by listing some of those that are engineering-related:

- Thorn + EMI
- RCA + General Electric
- ITT + CGE (France)
- Sperry + Burroughs

eems that UK was ahead of the game in bringing its major electrical-ing companies together under the GEC name, and other organisations have still to experience the process of digestion. General Electric of America has experienced this restructuring more recently, as described in the IEEE's *Spectrum* of February 1986. Cooke (1986) concludes that merger activity is rational, if it adds to the value of the enterprise.

Coales (1983) provides a thoughtful review of the role of the engineer in a large organisation. He shows that there is a long-term need for more engineers in the 'productive industries' throughout the world, and this need will exist whatever structural changes are made in organisation and finance. In a philosophical paper, Smith (1986) points the need for greater attention to industry through re-direction of education.

1.4 The concepts of hierarchy and functions

1.4.1 A pattern of change

A framework of organisation structure is desirable for any enterprise, if the situation of the nineteenth century Edward Lear's 'Jumblies' is to be avoided:

> In spite of all their friends could say,
> On a winter's morn, on a stormy day,
> In a sieve they went to sea!

It is necessary to plan and organise in order to achieve a task – the organisation establishment is not an end in itself. But when change is necessary because of changing needs, those involved may see the upheaval as unnecessary: to win their support they should be involved in the change process, through what is known as 'organisation development'.

In history we are aware that the Roman Legions were organised as an effective military-command hierarchy, that reached right across Europe, until the collapse of the central command. In medieval England there was a hierarchy of the Court, and of the baronial estates: this baronial style is still to be found in some family owned businesses.

During the 20th century the professional manager has assumed executive roles in most companies, and the Board of Directors can only be unseated with difficulty by the owners (shareholders). More frequently control of a company changes because another organisation makes an offer to shareholders for their shares, that they cannot afford to refuse. Thus changes tend not to be made to anticipate needs, but occur only after a serious decline in the company's fortunes. In Germany and Japan the banks play a large part in the provision of permanent capital, and exercise a closer supervisory role; so if a business is slipping questions are asked at a much earlier stage.

Life in some large corporations is not unlike that of a medieval court, if the Chief Executive has allowed this to be so. The courtiers will be sycophantic and practice gamesmanship, and to get on one may have to be a golfer or a chess player, with the right background and attributes. There have been a number of books in a cynical vein on 'how to do it'; for example, 'The corporate Prince – a handbook of administrative tactics' by Aquarius (1971), and 'Management and Machiavelli' by Jay (1969). The biographies of leading business people, about how they did it, are currently in vogue; for example that by Iacocca (1984). He typifies the 'American dream': the son of poor immigrant parents who makes it to the top, in this case rescuing Chrysler from bankruptcy on the way.

1.4.2 A professional approach to management

The professional approach to management owes much to Henri Fayol (1929), a French mining engineer who, in 1889 (at the age of 47) was appointed general manager of a large mining and chemical combine, a post he held for 30 years. He applied scientific method to his work and, in 1915, published his book 'Principles of general administration'. He presented an idealised model of administration as operating from the top down, with defined tasks for the people at each level of the hierarchy.

In general terms he defined the main functional groups appropriate to any business. This was an important step, for it presented 'management' as a universal process, whatever the nature of the business.

He saw the administrative process as consisting of five elements:

organising/co-ordinating/commanding/controlling/'looking-ahead'

His word for the last element was prévoyance, which included both forecasting and planning. Because he wrote in his own language, his work was not widely noticed until an English edition was published in 1929.

'Management' tends to be an Anglo-American usage, as it is still customary to use 'administration' in the Latin countries. The latter tend to have a more formal style, with limited discretion allowed to an executive, who is expected to administer according to the rules and instructions transmitted down from the top. Our more familiar use of 'manager' generally implies considerable delegated authority and discretion, provided that the person gets results.

The innovation that Fayol made was to present a 'model', albeit idealised, that could be taught, to illustrate how a operational organisation structure should work. He had the experience to recognise that the human factor would bring about deviation from the theoretical model.

This conceptual step enabled organisations to be purpose-built, and designed on the basis that 'form follows function', to satisfy the needs of the situation. Up to then establishments had usually just grown, under the influence of the personalities and aspirations of individuals.

Another contributor who was little noted at first was Chester Barnard (1968); his classic book 'The functions of the executive' having been reissued 30 years after it was first published. Formerly President of the New Jersey Bell Telephone Company, he made an analysis of the nature of business organisations, and was one of the first to highlight the contribution of informal organisation, and the moral aspects of leadership, so complementing Fayol's contributions to the more formal aspects. The story is brought up to date in 'Re-inventing the Corporation' by Naisbitt and Aburdene (1985), who have taken account of current trends in society and technology.

1.4.3 The pyramid hierarchy

The internal workings of a business were seen as a private matter in the early part of the 20th century and, where better methods were developed, they were thought of as secrets of the business: there was not much executive mobility, so ideas were only shared confidentially among small groups.

In the latter part of the 19th century, the United States developed some large-scale enterprises such as the railroads and the iron and steel industry. F.W. Taylor's innovations in work measurement provided the accountants with means of quantifying activities and separating out their costs. In Chandler and Salsbury's (1971) book we have the account of how Pierre Du Pont modernised an explosives and chemical business, already 100 years old. He transformed a cartel of small family businesses into the prototype of a centrally administered industrial company, following a policy of vertical integration to ensure continuity of supply of essential materials.

All this had been achieved before The First World War, and, afterwards, in 1920, the company found it necessary to assist General Motors Corporation in which Du Pont was a large investor. The automobile business had been hit by the post-War slump, and Du Pont supplied financial executives from their own strong team, to get GM's situation under control. Thus Du Pont's style of corporate administration, organisational structure and financial controls was carried over to the other company. It then gradually spread through the large coporations in the USA.

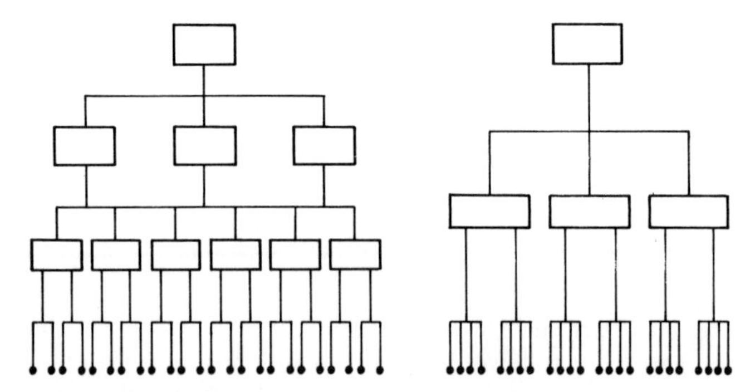

Fig. 1.1 *Reducing levels in hierarchy by increasing span of control and individual degrees of autonomy in operating units*

The story is taken up in the book by Sloan (1963), who was a key figure in General Motors until his death in 1966. He developed the system of autonomous trading Divisions, under a relatively small corporate top management, whose principal function was that of banker and initiator of any major new ventures.

The difference between the pyramid of the large corporation structure and that of the military command structure is this decoupling of the Divisions. The military chain of command is traditionally short, with few levels, which makes for speed of communication in action. A weakness in business is a tendency to add intermediate levels, such as by appointing deputies, so attentuating the up/down communication path. The span at each level can be increased (as sketched in Fig. 1.1) if people know what they have to do, provided that an effective management information system is available to signal problem areas.

Divisionalisation as practised in the first half of this century was no escape from bigness, as the industries that applied this form of organisation had technical reasons for operating large and indivisible plant. Delegation to smaller and more manageable units (100–300 strong) is a more recent development, and is associated mainly with the newer industries. In their 'Handbook of organisational design', Nystrom and Starbuck (1981) have made a comprehensive examination of the structure and consequences of organisational behaviour.

1.4.4 Rise and decline of line and staff

During the 1950s to 1970s it was usual to have a large corporate staff at the centre, undertaking strategic and long-term planning, and monitoring the performance of subsidiaries, sometimes by on-the-spot operational audits. This reduced the motivation and responsibility of the operational managers, yet the corporate staff were remote from contact with the market.

The 'line' and 'staff' pattern widely adopted was pioneered by the McKinsey Consulting Group: for a large organisation it typically provided two sets of functional Directors, as in Fig. 1.2.

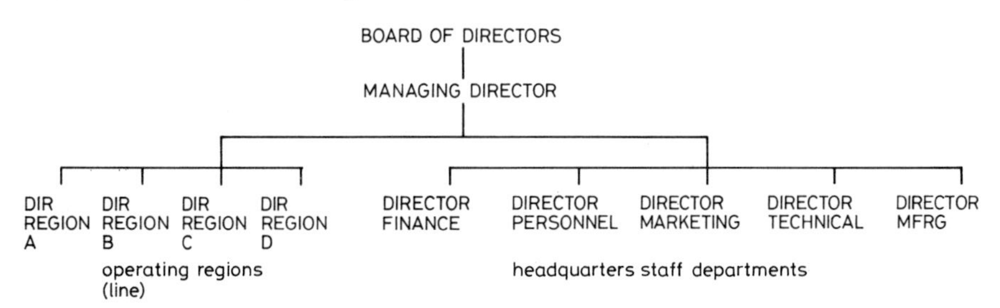

Fig. 1.2 ***'Line and staff' organisation structure***

Each operational region or division (A, B, C, . . .) would be the responsibility of a particular Director. Then at headquarters another set of Directors would each be responsible for a particular function: they would lay down practices and standards, carry out performance audits, and do the longer-term plans, rolling these forward annually. The operational directors (or managers) would control their own programmes for the current year, and collaborate with headquarters to determine budgets and plans for the following year.

The headquarters functions would be repeated in the operating units, but at a low level, executing the practices laid down. The style of headquarters control exercised in practice has ranged from heavy and detailed, to light and based on 'exception reporting'. Examples are the method that Geneen (1985) described in his autobiography, where the top team from each of ITT's subsidiaries gathered for several days each month at the Brussels headquarters. At GEC there has been a monthly review, but less elaborate, and with a much simpler reporting system (Lawless, 1986).

The trend in the 1980s is to reduce drastically the corporate headquarters staff, and put more responsibility directly upon the management of each operating unit. It can be argued that this is a natural progression, and that operational line managements were not ready for this responsibility a decade or more ago, and that more staff guidance was then essential.

Morley (1986) has described the impact of privatization on British Telecom, in its transition from the former Government Department, the General Post Office.

1.4.5 Downward command versus upward concensus

Great concern was expressed on realisation in the 1970s and 1980s of the relative success of Japanese companies. They achieved this result by making good use of their own cultural disciplines, and by carefully choosing the most effective Western techniques and practices, and adapting them. What is seen now are the results, but the process has been going on for some 30 years or more.

A great deal has been written by way of explanation, and a good analysis is by Livingstone (1984), who described a year spent in Japan on a staff exchange between British Telecom and a Japanese company. As is now well known their style of management is that, within broad guidelines, there is upward communication of concensus from the personnel as a whole, linked also horizontally with

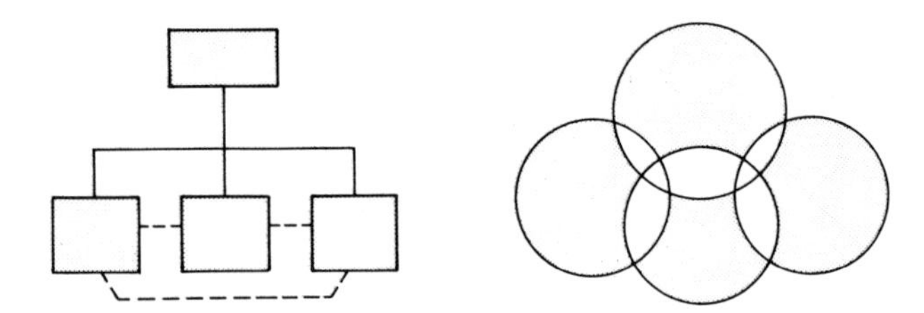

Fig. 1.3 ***Western organisational 'boxes' and Japanese diffused pattern***

associated groups. This ensures that everyone is well motivated. Fig. 1.3 is an attempt to illustrate the differences. The Western tendency has been to have well defined organisation 'boxes', with limited communication, mainly from above: the Japanese relationships diffuse together.

Recapitulating the styles of management already mentioned:

- Roman legions, Governors, and Senate
- Medieval courts
- Baronial estates
- Family-owned businesses
- Limited liability companies

- Model functional structures, designed to suit needs
- Du Pont and GM's pioneer decentralisation
- Holding companies, and subsidiary companies
- Military command structures
- Central staff and operational line management
- Formal structure versus diffused concensus

It is stimulating to conjecture about the future trends in this progression, which will be influenced by at least three factors: market forces, technological changes, and managers who are more knowledgeable.

1.4.6 Integrated networks and automated manufacture

The diffusion in the Japanese model is through the interaction of people, and we can now expect further diffusion of functions through the interaction of currently available technologies, although its spread may not be rapid. The Policy Studies Institute surveyed UK factories and determined that the proportion using microelectronics was 10% in 1979, 20% in 1981, and would pass 50% in 1986. This is a crude measure, but it indicates that the potential for full integration of communication networks will come only gradually.

However, there is the potential technical ability to integrate all functions in an organisation into a single operating system, and the relationship between functions

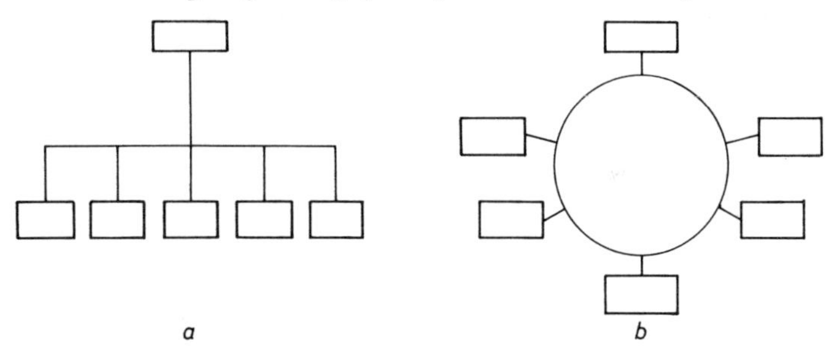

Fig. 1.4 a *Conventional organisation structure*
b *Departments linked by local area network (LAN)*

will then be more like (*b*) in Fig. 1.4 than the conventional (*a*) structure. People will access the system to perform tasks that are their responsibility, but are likely to be more flexible and cover a wider proportion of the functions. To avoid confusion it will still be very necessary to be clear who does what.

As in any new area of activity, the terminology is not yet fully matured, and the following are among the principal aspects:

Communication networks

TOP : Technical office protocol (Boeing) linking offices, departments, suppliers, customers

CNMA : Communications networking for manufacturing applications (European)

SPAG : Standards promotion and awareness group (GEC and European companies)

OSI : Open systems interconnect (IBM)
SNA : Systems network architecture (IBM)

Automated manufacture

AMT : Advanced manufacturing technology

General title, embracing the following:

FMS : Flexible manufacturing systems
CIM : Computer-integrated Manufacturing (IBM)
MAP : Manufacturing automation protocol (General Motors)
Motorola have developed a chip-size interface controller
CAD : Computer-aided design
CAE : Computer-aided engineering
CAM : Computer-aided manufacture

MAP is a set of computer protocols, for multivendor manufacturing automation systems, combining the GM standards and the ISO 7-layer system (open system interconnect) – it is described by Moore (1986).

Alternative protocols exist, sponsored by the originating companies, and there are several initiatives directed towards securing compatibility, including that by the International Standards Organisation (ISO). Although the initial choice is difficult for a first-time user, interfacing between users on different protocols is a valid option as the cost of the necessary power for the data conversion will continue to fall.

Typical steps in a strategy for introducing AMT would be along these lines:

- set up a task force, with all main company and external functions represented
- reinforce skills with specialist consultants, as necessary
- outline needs of an overall system, appropriate to the firm's business, as a non-specific 'requirement specification'
- identify each element of a model system to cover 100% of functions
- evaluate each element, and rank in order of cost effectiveness
- plan to introduce first the elements with best pay-back
- make a long-term plan for working towards a 100% system
- review progress at intervals and re-evaluate next step
- determine where to stop short of 100%, in relation to flexibility for future configuration changes of 'islands of automation'

An example of a project along these lines is described by Beaty (1986). Kaminsky (1986) describes MAP implementations and Farowich (1986) outlines TOP. Shortland and Prince (1985) review AMT overall.

The history of automation systems is of some interest. 'Automatic manufacture' was developed by production engineers in the 1930s, in particular with purpose-built machine tools, linked by transfer mechanisms, mainly to produce car engine and power train components. This stemmed from the early development of the mass production of interchangeable parts. There were isolated examples in the early 19th century for firearms manufacture (in France and America), Bramah's lock-making machinery, and the Portsmouth rope block-making machines described by Gilbert (1965).

These systems were highly productive, but very inflexible, as all tooling was single-purpose. Later some flexibility was obtained with simple programming of machines, at first with plug-board selection, and then with punched-tape and magnetic-tape controls. In the early 1940s there was a pioneer system for fabricating integrated (but not miniaturised) circuits for complete radios, the ECME machine of John Sargrove (1947). The 'state of the art' some 30 years ago is summed up in books by Lilley (1957) and in two Conferences arranged by the Institution of Production Engineers (1955 and 1957), and in a review by DSIR (1956). The progress made since is mainly in the application of much more advanced data processing to the programming and control, as foreshadowed by Lord Bagrit (1965).

1.4.7 Flexibility, rapid response and new work practices

Traditional engineering manufacture was inflexible because the cost and time required for retooling was high, whenever an additional or different product was required. There was much emphasis on the economics of scale: companies concentrated on volume and continuity of production, rather than be led by market needs.

There was a tendency to build up large stocks of finished product before taking a decision to adjust the volume of production and, in turn, to carry large quantities of materials and components.

At times of high interest rates the financing of these stocks could cripple a company, and result in the attrition of its working capital. A sudden cut-back of production would send a ripple of dislocation through the network of suppliers. The economic disturbances (induced oscillation) caused by 'stop-go' behaviour by large producers was ripe for systems analysis and the application of 'proportional control', to give more rapid feedback of control signals.

In the textile business this phenomenon was known as the 'trade cycle', and was accentuated by the network of several process stages of spinning, weaving, finishing, wholesaling, making-up and distributing, plus the stocking at the retail level, and the effects of the seasons and of fashion. The problem was accepted as being the nature of things until the 1950s, when vertical integration of businesses in the chain rationalised and reduced the number of decision points, and the Government began to publish Trade Statistics, which made generally known the levels of industrial stocks.

Other industries had similar problems, with the same consequences of 'feast and famine', economically inefficient, and the cause of distress to the workforce during the bad times. In the 1970s the Japanese became known for the 'just in time' concept where, by tight scheduling of supplies, very little was in the pipeline between material suppliers, component subcontractors, and the main assembly plant.

This was no new idea – it was used by Henry Ford in the economic depression that started in 1920 when, by strict discipline over costs and in the scheduling of supplies, and by drawing down buffer stocks and material in the pipeline, the company averted a cash-flow crisis. Simonds (1946) records that in 1920–21, Fords were due to pay $18 million in Federal income tax, at a time when they had 93 000 completed cars in stock. Various measures were taken, including reducing the overall production cycle (raw-material/finished-car) from 21 working days to 14. This released $28 million by converting stock and work in progress into cash, which saved the company from bankruptcy.

This was achieved by intensive use of staff in progressing, expediting, liaising with suppliers, and physically organising a more rapid flow of materials. When the crisis was over this tightness of control was relaxed somewhat, the labour costs being seen then as more than the expense of financing buffer stocks.

With today's technology, it is economic to use communication links to make distant independent suppliers effectively part of a single production line, provided that they will accept that they must work to these disciplines, and are virtually 100% reliable.

There is a continuing change in the relationship between labour cost per hour, and the related capital cost, or average 'investment per head' and, in comparing the performance of companies, criteria such as these are revealing:

- capital employed per employee
- sales value per employee
- added value per employee
- 'stock-turn' (rate at which stocks are used)

As the degree of automation is increased, nearly all the direct labour is taken out of the process, and the real labour cost is that of all the support staff who keep the system running, which in simplistic accounting was regarded as 'overhead'.

The more sophisticated applications of communication networks and automated manufacture are very costly in capital investment. In a competitive world, the organisations that operate these resources most intensively will have the advantage in costs. To justify the capital commitment it must be recovered in fewer years, which is necessary anyway to afford replacements and more frequent updating. This is in contrast to traditional engineering factories, where single-purpose major machine tools 50 years old could be found in regular use.

There are 168 hours in each week, and allowing for maintenance and down-time it is quite feasible to plan for 120 hours utilisation. The co-operation of staff is necessary to achieve this coverage, but with 'early' and 'late' shift working, hours need not be excessively 'unsocial'. With modern robot equipment it is quite feasible to let a night shift run unmanned, provided that there is automatic shut-down if any problem is detected by the monitoring system. Although relatively new in general manufacturing, continuous working has long been practised in some industries such as glassworks, steel and petrochemicals.

Developments in AMT are co-ordinated at the international and national levels, it being necessary both to work towards compatible languages and protocols, and to stimulate adoption of the new technologies in the industries where they are most needed.

In the United Kingdom, the Department of Trade & Industry is active in this role, in conjunction with the National Economic Development Office (NEDO). The National Economic Development Council is UK's forum for economic consultation between government, management and the unions.

The Engineering Institutions have established the Engineering Manufacturing Forum, which arranges interdisciplinary lectures with authorative speakers: it represents the Production, Mechanical, and Electrical Institutions. This theme of collaboration, and the need to develop the national manufacturing capability, was reviewed by Banks (1983) of BICC. A survey sponsored by BIM in 1986 showed only moderate progress. Some two-thirds of companies that probably could make some use of CAD/CAM had done so, but half had not yet seen significant gains. Of 250 manufacturing plants, 64 had experimented with FMS technology, but two-thirds of these reported low payoff. A poll of future intentions showed about 50% put emphasis on CAD/CAM, 25% on FMS and only 16% on robotics. In summarising these results, Professor Colin New of Cranfield School of Management comments that these attitudes compare unfavourably with the West Germans, French, Japanese and Americans.

1.5 Back to smaller units

1.5.1 Smaller and simpler

Schumacher (1973) was chief economist and head of planning at the British Coal Board: he thought deeply about alternative societies and the problems of the Third World. He concluded that they should reject imitations of Western models in order to satisfy criteria not based entirely on economic considerations. In his book 'Small is beautiful', which eventually sold over a million copies, he coined the term 'intermediate technology', meaning a level that is more effective than primitive methods, but not requiring great capital investment.

He wrote that any third-rate engineer or researcher can increase complexity; but it takes a certain flair of real insight to make things simple again.

His message of how to help the Third World to help itself gradually won acceptance, and this reinforced a growing recognition in the developed countries that unnecessary complexity is to be avoided.

For example, an interlaced metropolitan railway ('rapid transit system') with many track cross-over points, is a signalling engineer's nightmare. But if separate branches run into a transfer station, where passengers can cross a platform to join the other line, control of the system is greatly simplified, and track maintenance is reduced.

There is usually a more simple way to do things: for example a complex project can be made easier to manage by setting up secondary projects within a simplified primary critical path network (Fig. 1.5).

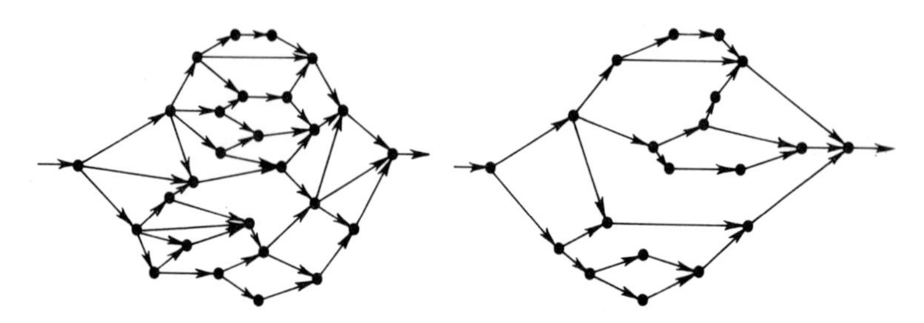

Fig. 1.5 *Complex project critical path network is simplified into a smaller number of secondary projects*

1.5.2 Dismantling the large groups

Large companies have been inclined in recent years to dismantle their large headquarters corporate establishments, dispose of traditional prestige headquarters buildings, and push responsibilities for tasks such as forward planning, personnel and marketing back to where the action is, at the operational sites. GEC disposed of Magnet House in Kingsway (as have Kodak, who were neighbours). The amalgamation of the other electrical engineering companies accounted for the disappearance of AEI's HQ in Grosvenor Place, and those of BTH and English Electric's around Aldwych: what remains now is a modest address at Stanhope Place, accommodating only 100 people, of a Group staff of nearly 200 000, worldwide.

To Lord Weinstock must go credit for demonstrating how to run a large Group (turnover of the order of £5 billion), by pushing responsibility for results down to a federation of some 200 tightly controlled small businesses, of which the majority of managers are engineers by background. The tightness may have been uncomfortable at times for these managers, but it did produce annual profits above £700 million in the mid-1980s, an outstanding performance on world scale in this industry. Net income as a percentage of sales is twice that of AEG–Telefunken, two or three times better than comparable Japanese companies, Matsushita Electric, Hitachi and Mitsubishi, nearly ten times that of CGE (France), four times that of Philips and of Siemens, while Thomson (France), Montedison (Italy), ASEA (Sweden) and Brown Boveri (Swiss) had net income close to zero.

Unlike the technique that Geneen (1983) describes in his book about his time at ITT, the GEC approach is relatively simple, as Turner (1985) has observed.

One of Bernard Shaw's sayings was:

> . . . the golden rule is that there are no golden rules

and what may work well in one company culture may be a disaster elsewhere, but there is no doubt that simplification wins.

A traditionally complex hierarchy produces a tall organisation pyramid, with many high-ranking staff people who are not directly productive (and may even get in the way of 'operations'). They will mostly be engaged in liaison and attempting to overcome the communication problems caused by many layers of management, and a wide span of operations.

In the unsettled industry of microelectronics, Motorola has been one of the most consistently successful firms, and O'Toole (1985) attributes this to the grouping of some 60 000 employees in teams of only 50–250 each. There is a high degree of worker involvement, and well developed upwards and downwards communication.

A statement by Racal (1980) said:

> It has been Racal's policy since the early 1960s to organise itself on the basis of small and medium-sized companies which operate as autonomous profit centres. . . A key factor is that the directors, the sales, engineering and production activities and other departments should be close to each other and capable of daily contact. This creates an environment of quick reaction to customers, sensible product planning and speedy solution of problems.

Racal's profit ratio is good by comparison with the other companies mentioned, being about two-thirds of GEC's, and it is about a quarter GEC's size.

1.5.3 Intrapreneurs

Large Groups have attempted to gain the advantages that small units demonstrate, by setting up independent enterprises, with the intention that they should be entrepreneurial is management style, while remaining wholy owned: hence the term 'intrapreneur' used for the book by Pinchot (1985).

The 3M Company has done this effectively over many years, as a marketing method for products of specialised end use. This management formula can work well for a new product in a new market, that is disconnected from the Group's main business. IBM launched the small PC computer in this way, contracting externally both for hardware and software.

One benefit is that those opportunities serve to retain ambitious people, who otherwise would be likely to seek opportunities elsewhere, and be lost to the Group.

1.5.4 Buying-in and buying-out

There is a growing tendency for holding companies and large Groups to both buy and sell segments of their businesses, in the same way that an investment group would change its portfolio of properties or shares, from time to time.

A young entrepreneurial company may have developed a unique product, and have found a promising market niche, missed by the large companies. If it is a field that they wish to enter, it is often simpler for them to buy up the young company, than to set up a company in opposition. The original entrepreneurial team usually depart sooner or later, rewarded not so much by golden handshakes, as by the capital they receive for their equity in their company. Hopefully, they may succeed in repeating the process again with another new venture, unless their good fortune with the first was mainly attributable to being in the right place at the right time. The successes tend to be publicised as models, but for every one of these there will be a proportion of failures, and a middle group of ventures that never really 'take off'.

The reverse process is a 'management buy-out'. The parent Group wishes to either close down or sell a unit. If the management team affected are able to secure the assistance of a merchant bank, they can be in a position to make an offer. A typical 'gearing' is that they must find 10% or more of the money from their personal resources – they usually have to mortgage their homes, and their banking supporters will favour this because, as Samuel Johnson remarked:

> 'when a man knows he is to be hanged in a fortnight, it concentrates his mind wonderfully'

The level of motivation of the buy-out team is likely to be higher in their new role as owners, so usually their efforts generate better results than before, helped by reduction in overheads and some rationalisation of the activities.

There have been several recent books which are helpful, including those from EIU (1982), McLachlan (1983), Webb (1985) and Wright (1985). Some Merchant Banks and firms of accountants will provide advice: for example booklets on 'buy-outs' are available from Spicer and Pegler, and from Peat Marwick. A Management Buy-out Association has been formed.

McLachlan's book, and a paper by Sir Peter Thompson (1985) describe the denationalisation and buy-out of the National Freight Corporation. This is a very large operation with 10 000 employees, and some £50 million had to be raised. By skilful direction from the top team, the many difficulties were overcome, and a good proportion of the employees became shareholders, and have seen a satisfying increase in the value of their investment. Despite its size, NFC has many of the characteristics of the smaller companies, as most of their people are attached to small or medium sized depots, each of which has the atmosphere of a small firm. This is also an example of a 'share ownership' scheme, although the effectiveness of these in established companies may be in doubt, as the employee is tending to 'put all his eggs in one basket': such schemes are described by Copeman (1984).

1.5.5 Positive discrimination in favour of small units

There is a general feeling among economists that small entrepreneurial firms are likely to be the main source of growth in the economy, and the current mood is to provide preferential assistance, not available to large firms, as in these examples:

- DTI operate a number of technical assistance schemes, which are available to companies with less than 500 employees
- The National Computing Centre (NCC, Manchester), in association with Chambers of Commerce, can advise the small-business sector on the choice of appropriate business software packages and the machines to run them on
- The Treasury may introduce tax concessions, in return for agreement to link pay to the fortunes of a company
- An expansion is likely in profit-sharing schemes, with tax breaks for approved schemes: but this does also imply loss-sharing
- Agreed total annual hours for employees, but with flexibility to absorb peak and slack periods

- Special facilities and services, such as 'starter-units' of business accommodation (see Section 6.7)
- Government Agencies, when placing major contracts, will require the company to place a proportion of its subcontracts with very small businesses.
- Government Departments will adopt simplified procedures for small businesses, for example, VAT accounting.

Much was learned in the 1960s from attempts to re-establish ailing companies as co-operatives. The cards were usually stacked against success, as there were already marketing problems, and those in charge had limited experience. There were some idealistic attempts to manage by concensus of a committee, and eliminate 'managers', but this is not very effective when day-to-day decisions are necessary. In most small companies, working Directors may be unclear when their role is that of a member of the Board, and when it is appropriate to follow their role as an executive manager. The Institute of Directors can assist its members, many of whom are in the smaller companies, and a number of useful publications are available, together with an advisory service.

1.6 Objectives, reviews and performance measures

1.6.1 Establishing dynamic objectives

In history there have been static periods of apparent stability, where there was negligible growth or change, and a generation of job holders were then unready for change when it came.

The process of establishing objectives assumes dynamic environments in industry, technology, commerce, public services and socially. The process consists of establishing agreement on a broad statement of the 'mission', charter, target, objective or goal, and then working towards it by iterative steps of review and measurement of performance.

Different meanings and usage will be found for these words – 'mission' is widely used in the USA (from its military origin), and it does imply the spirit of dedication that distinguishes some of the most effective enterprises.

Humble (1965) sees the dynamic process as circular flow (Fig. 1.6), and this presentation tends to de-emphasise the superior/inferior implications of communication down and up a hierarchy. Various side activities are spun off from the continuing cycle of review; for example, the need for, and results of, management development. In Fig. 1.7, Humble's circle diagram (*a*) is compared with Black's (1934) classic diagram of the feedback process.

Two other books edited by Humble (1970 and 1973) contain accounts by people who have introduced systems of objectives in their own organisations. For example, Hawthorne in the 1970 book describes the use of objectives in R & D. Odiorne (1979) makes a more recent assessment of the way that 'management by objectives' (MbO) has been used and misused, and provides this definition:

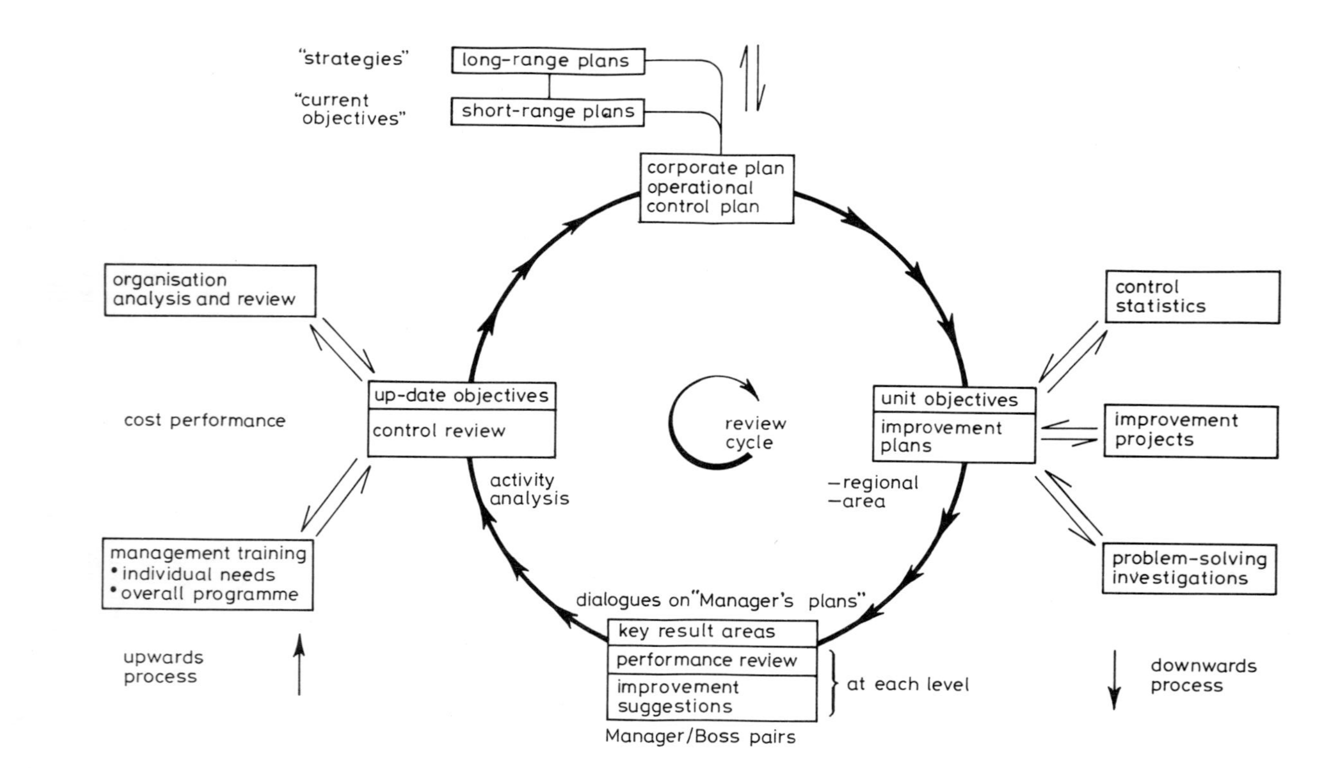

Fig. 1.6 *Dynamic process of reviewing objectives, and spin-off of side activities from the cycle*

A process whereby the superior and subordinate managers of an organisation jointly identify its common goals, define each individual's major areas of responsibility in terms of the results expected, and use these measures as guides for operating the unit and assessing the contribution of each of its members.

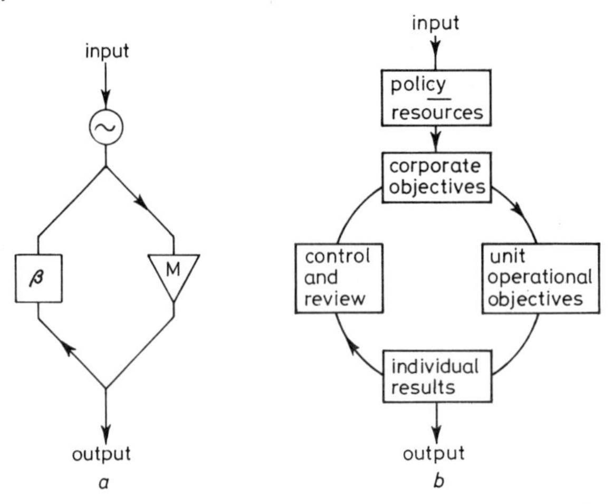

Fig. 1.7 a *Circuit diagram (from Black, H.S.: 'Feedback Amplifiers', Bell System Technical Journal, Jan. 1934)*
b *Dynamic nature of system of management by objectives (from Humble, J.W.: (BIM, 1965))*

1.6.2 Feedback of corporate performance

Leaving until later pages the human and personnel management related aspects of the review process (Sections 1.8 and 2.4), the corporate need is for prompt quantified feedback of results, performance and achievement, compared with the plan or programme. If on target, no managerial action is necessary, but there will be corrective action if variance is appreciable; 'performance' involves more than just the processing of financial information (which is described in Section 3.6, and illustrated in Fig. 3.2). The process is most clear-cut in a project or task-force situation, where intermediate targets have to be met; in particular those on a critical path of the planning network.

'Too much' action may demand attention, as 'too little' will. In the project situation there is no point in putting resources into an activity when there is already plenty of 'floating' time, if those resources can be devoted to something currently more urgent. In manufacturing, overproduction results in excess and unbalanced stocks, representing frozen capital resources.

A 'task force' has similar characteristics, but is directed towards diagnosis and improvement of a situation, working towards a clearer definition of objectives, and the formulation of plans for improvements or the development of resources. Pascoe (1970) describes the methodology, and its application in a large engineering stores and supply organisation, where it was used to analyse information from areas identified for major improvements.

1.6.3 Performance measures, and the concept of continuous improvement

In one of his books, Drucker (1974, chapter 13) describes how, in 1910, Theodore Vail defined the mission and business objectives of the Bell Telephone System (AT & T) as: 'our business is service'. He set up 'customer satisfaction' standards, and measures of performance and results. Over a period of more than 50 years Bell both improved its standards and reduced its rates. Drucker comments (page 478) that, without those tangible yardsticks, the statement of intent would have been only a pious hope.

Tomlinson (1982 and 1983) has said that neither quality nor productivity is easy to define in a public telecommunications network. British Telecom has since the early 1960s operated internally a comprehensive structure of performance measures and, until recently, published a selection, for example (at March 1982):

calls connected successfully (local):		63·3%
reasons for remainder	no reply:	28·4%
	due customer:	7·0%
	due BT:	1·4%
yearly faults per telephone:		0·58%
faults cleared by end next working day:		79·6%
Operator service:		
calls connected within 15 s:		87·4%

Now that the Office of Telecommunications (OFTEL) is established, it intends to publish a series of measures. The Director General, Prof. B. Carsberg (1986), has explained that charges for service are expected to conform to the rule ($RPI - X$): this is seen as an efficient method of regulation because it is inexpensive to administer, and it protects consumers by ensuring that the annual change for a 'basket' of inland tariffs will be X percentage points below the change in retail price index; i.e. less rapid than inflation. The indicated value of X is 3%; so cost effectiveness has to improve on average at this annual rate. This is feasible in a high-technology industry, but another value of X would be necessary in a different industry.

Tomlinson (1983) published Fig. 1.8*a*, which shows the considerable improvement achieved in fault rates over 15 years, much of which is due to new technology. Fig. 1.8*b* shows a factor-of-two improvement in productivity, in terms of telephones per employee, which he attributes to:

- improved technology
- substitution of capital investment for labour
- working method changes
- computerisation
- management and staff motivation
- manpower planning

This performance is better than some countries, but is below that in the USA and Sweden. These national differences may be due to a variety of factors, but Prof.

Eilon (1986) stresses, and Tomlinson reiterates, that it is important to focus on the rate of incremental improvement, and the most effective ways of bringing this about. For services such as this there is a clear relationship in quality between performance and cost: in this case, the tendency is to 'peg' performance, in order that productivity gains will offset inflation.

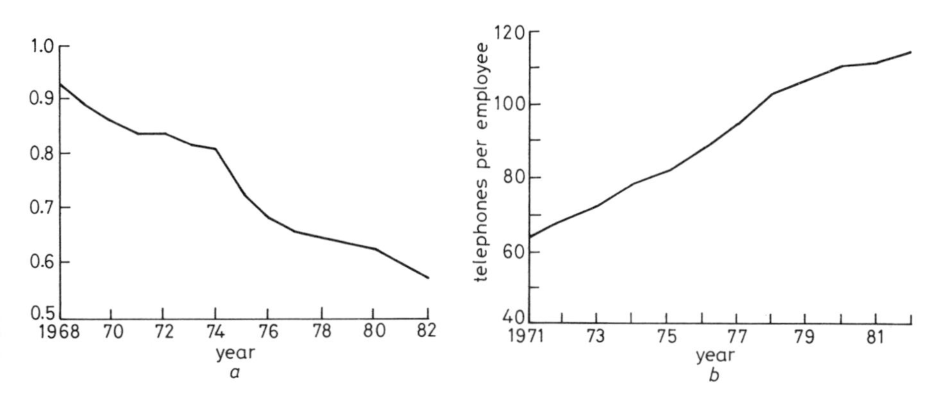

Fig. 1.8 a *Yearly fault reports per telephone*
b *Telephones per British Telecom employee (from Tomlinson, 1983)*

Other service organisations have adopted similar measures, and in Chapter 2 (2.4.4) examples are given of the application in Mullard's Marketing Department. The RAC now promise to reach 8 out of 10 motorists within one hour of being called, and the Post Office sets targets for delivery performance on first- and second-class letters. British Rail's 1985 Code of Practice has the targets of 85% of Intercity and 90% of other trains arriving within 5 min of scheduled time.

A full scheme of measures has to be more complex than these examples. A control is necessary on the residual percentage to avoid a 'long tail' of unsatisfied exceptional requests. A typical control schedule might be:

80% of faults cleared by end of next working day
95% by 3rd working day
100% within calendar month

In order to satisfy such targets, a well thought out control system is required. The Americans use the term 'dispatcher' for a controller at a central point who schedules and monitors the deployment of the resources available, and deals with priorities and unusual situations.

1.6.4 Examples of Codes of Practice

The trend is away from control by legislation, towards 'self control' in business, but with provision for a 'watchdog' function as a channel for responding to consumers' interests.

An organisation (public or commercial) will issue a Code of Practice describing how it intends to do business, possibly drafted in conjunction with an advisory group of its clientele. There will be some channel of appeal laid down, either through the public Ombudsman, or a specially established Office, such as OFTEL, or the Office of Fair Trading, or the National Consumer Council.

The telephone, postal, gas and electricity services issue Codes of Practice. The foreword introduction in British Telecom's document says:

> This Code of Practice explains many of the services offered, sets out our commitment to our customers, and tells you where and how to get advice and help

The background was explained in an article by Marcus (1979), and the Code now is included in the back of telephone books.

British Airways has to compete internationally, using equipment no different to their competitors: its statement of missions and goals (1986) is: 'To be the best, safest and most successful company in travel, tourism and transport', and sets targets for cash flow and to earn 'good profits to satisfy shareholders'. Other goals are to be known as 'the most efficient and most customer concerned airline, and for the name to be synonymous with people concern, high achievement and general business ability, and to improve as a good employer'.

It may create problems to aim to be 'the best': IBM covers this, and the inter-country differences, by simply aiming to provide employment conditions 'among the best' in each territory where it operates.

In any of the services to the public, staff training in public relations is important, because people do not behave rationally, particularly if in unfamiliar surroundings, and a practical understanding of their psychology, and a proper motivation, is necessary for all staff. Many large organisations now appreciate that their image in the eyes of the general public is only as good or bad as the impression left by the last employee/client contact.

Most organisations will accept back or exchange returned goods without demur, while operating simple internal checks against abuse of the system. In consumer goods the businesses that operate these policies seem to have had the most success. An article by Lele and Karmarkar (1983) examined customer expectations, and cost-effective ways of satisfying them.

In the function of purchasing, the Institute of Purchasing & Supply has established a code of ethics. A Code of Practice sponsored by the Department of Employment exerts pressure for timely trade payments to smaller businesses, by their large customers: in this instance the Government have said that if the Code fails to make an impact they will introduce legislation.

1.7 Corporate and strategic planning

1.7.1 The value of foresight

The problem facing most large companies was summed up by Derek E. Taylor (1979):

> . . . many businesses are now too big and/or too complex either to be understood by any of their employees or to be managed easily. Companies have

> grown to a size where their managers have become so preoccupied with departmental issues as to lose sight of the business as a whole. Often it is only the Chief Executive who has an oversight of the 'whole business' and he may be too involved with the problems of managing the organisation to make the quality of business decisions demanded by today's competitive environment.

At the other end of the scale, small businesses are either struggling to survive on a day-to-day basis or, if more fortunate, they are much too busy with current projects to look further ahead.

Perrin, one of the British pioneers of strategic planning, has said:

> Those in management who hold prime responsibility for looking to the future must see the danger of having tunnel vision and setting goals which are too low. In order to survive, they must begin with the premise that the high technology industry of the future must have a global perspective.
>
> Perrin (1985)

Much of the definitive work on business strategy formulation has been by H. Igor Ansoff of the Carnegie Institute of Technology (e.g. 1984, 1979 and 1968). Earlier, he was with the RAND Corporation, and with the Electronics Division of Lockheed.

In the 1950s, corporate and strategic planning was coming into use in the UK, and was successful in some of the larger companies during the 1960s, a period of relative economic stability and growth.

The Strategic Planning Society (as it is now known) was established in 1967, and has over 1000 members plus about 150 corporate subscribers. The 1970s proved exacting for this new profession, because the earlier techniques were taxed by increased rates of inflation, intensified market competition, and a heightened degree of uncertainty. The longer-term projections were proving to be erroneous, due to such step changes as the pressure by OPEC on oil prices, which caused a ripple effect on markets and world economics, a situation foreshadowed by Drucker (1971) in his book 'The age of discontinuity'.

In order that strategic planning may be more effective, the trend now is to push the planning task downwards to the managers who will implement the agreed corporate strategies, in contrast to its earlier position as entirely a staff function at Group headquarters. A version of the process is described by Malpas (1985), Managing Director of British Petroleum, and formerly with ICI.

Strategic reviews are now made more frequently and at shorter range, except where there are special circumstances. The operational plan can be adjusted quickly as conditions change. The direct involvement of the operational management has a three-fold advantage:

- those with intimate knowledge of the market and/or the technology tend to have an intuitive feel for the likely events
- by being involved in policy formulation, they are commited and motivated to implement the agreed action programmes
- as further changes are identified, they can be responded to quickly, via a tight internal feedback loop

Long-term strategies remain necessary for those projects or markets that take a long time to construct or develop, such as power stations, mineral and petroleum extraction and telecommunications systems. The provision of power-generation capacity tends to overshoot demand, but this does allow the withdrawal of less efficient units, and fortuitously provides cover for delays in commissioning. In the business of mineral exploitation, projects proceed in stages, and then may go onto the 'back burner' until the major investment in construction is justifiable. O'Hara's (1985) paper on British Telecom's planning refers to the need to look to the year 2000 and beyond.

1.7.2 How to do it

The role of the corporate planning staff in a large organisation is to provide the framework for integrating the contributions of the several specialist disciplines and operating units. The planners will have expertise in techniques, and knowledge of how to access external sources of economic and marketing data. A comprehensive review of these sources is given by Taylor and Redwood (1982): having been made aware of these sources, it is only necessary to secure the latest figures.

The strategy of an enterprise has to take into account and balance the influence of many inputs, in particular:

- the organisation's declared purpose, mission, objectives and goals
- current plans and preferred future objectives of the several operating units
- inputs of market intelligence, with particular reference to:
  - perceived strengths and weaknesses
  - identifiable opportunities and threats
  - intentional or fortuitous availability of market niches, and products with a market edge
- expectations of the several 'stakeholders' in the enterprise:
  - obligations to regular clients
  - employees
  - investors
  - external community, local and national

More specifically, the marketing function will develop its own preferred strategies for markets and products, and their views will carry more weight in a market-led company than in one where the main concern at policy making level is with technology and manufacture. In a medium-sized company, without a specialist corporate planning staff, the process can be carried out adequately by part-time working parties. These may be made up from a cross-section of representatives of the several functions.

A single working group is appropriate in a small company or, in a more complex situation, the work can be shared through a two-tier structure. Each of several groups can handle a main aspect, and they will report to a policy making group at Director level. John Argenti (1984) has developed a 'do it yourself' system for use by smaller firms.

Such informal arrangements provide opportunities for the involvement of a range of staff at all levels. When the corporate plan is rolled forward annually, the groups can be partially reconstituted, to bring in fresh contributors.

Useful books are those by Fawn and Cox (1985), a joint publication of the ICMA and the SPS, that by Hussey (1978), together with the paper by Hendry (1986).

The contributors to study groups will throw up a number of apparently viable options. These will have to be evaluated quantitatively by modelling, using the methods of operational research and statistical analysis, and by the development of alternative scenarios.

A great deal of useful literature is available, particularly in *Long Range Planning*, the journal of the Strategic Planning Society, and in the *Journal of the Operational Research Society* which has been published since 1948; also in Wiley's *Journal of Forecasting* and Butterworth's *Futures – A journal of forecasting and planning*.

The OR Society has a membership of 3000, and it defines OR as:

> The application of the methods of science to complex problems arising in the direction and management of large systems of men, machines, materials and money in industry, business, government and defence. The distinctive approach is to develop a scientific model of the system, incorporating measurements of factors such as chance and risk, with which to predict and compare the outcomes of alternative decisions, strategies or controls. The purpose is to help management determine its policy and actions scientifically.

There is an active International Federation or Operational Research Societies (IFORS) that publishes an excellent series of Abstracts, which include case studies related to engineering management.

1.7.3 Examples of planning

There follows a short list of descriptions of how representative organisations set about their corporate and strategic planning (the journal *Long Range Planning* is abbreviated *LRP*):

British Airports Authority
TURNER, D.: *LRP*, Oct. 1985, **19**, pp. 49–54

General Electric Company
TURNER, G.: *LRP*, Febr. 1985, **18**, pp. 12–18; also LAWLESS, J.: *Business*, April 1986, **1**, p. 79

GenRad Inc, Waltham, Mass.
CRAIG, S.R.: *LRP*, April 1986, **19**, pp. 50–56

Guest Keen and Nettlefolds
TURNER, G.: *LRP*, Oct. 1984, **17**, pp. 12–16

Imperial Chemical Industries
TURNER, G.: *LRP*, Dec. 1984, **17**, pp. 12–16

International Computers Limited
MARWOOD, D.C.L.: *LRP*, April 1985, **18**, pp. 10–21

N V Philips Gloeilampenfabricken, Eindhoven
DEKKER, W.: *LRP*, April 1986, **19**, pp. 31–37
Shell Nederland B V
LEEMBUIS, J.P.: *LRP*, April 1985, **18**, pp. 30–37

Also, more general descriptions:

CALDECOTE, VISCOUNT: 'Competitive position of European Telecommunication Industry', *IEE Proc.*, 1986, **133**, Pt. A pp. 365–368
CONSTABLE, J.: 'Diversification as a factor in UK industrial strategy', *LRP*, Feb. 1986, **19**, pp. 52–60
GREEN, G.J.L. and JONES, E.G.: 'Strategic Management Step by Step', *LRP*, May 1982, **15**, pp. 61–70
HOULDEN, B.T.: 'Survival of the Corporate Planner', *LRP*, Oct. 1985, **18**, pp. 49–54 (based on a survey of 105 corporate planning units in the UK)
WALTECH, P.: 'US Semiconductor industry: getting it together', *IEEE Spectrum*, April 1986, **23**, pp. 75–78 (despite a history of competition, US manufacturers are co-operating in research to gain an edge in the international market place)

1.7.4 Evolutionary change

The corporate planner has a relatively easy time where change is evolutionary and continuous. He can extrapolate, develop 'best' and 'worst' cases, and accept controlled risks using 'intercept technology' and the 'learning curve effect' (referred to in Chapter 5 (5.4).

However, he is effectively gambling with statistical probabilities, and not all the bets can be hedged. Forecasts based on different models give different answers.

For example, even 15 years after the shock of the increase in oil prices by the Organisation of Petroleum Exporting Countries (OPEC), residual fluctuations continue: a downward trend caused the Treasury to revise its estimate of the UK balance of payments current account surplus:

from £4·0 billion, as estimated in November 1985
to £3·5 billion, in March 1986

Forecasts for the UK economy are made by many organisations, and there follows a selection of their estimated percentage increases in the gross domestic product over the figure of 12 months earlier:

	1986	1987
Treasury	3·0	2·5
FT average (of 25)	2·5	2·5
Goldman Sachs	1·9	3·1
National Institute	1·9	1·7
Liverpool University	3·5	3·1
London Business School	2·4	2·9

Even short-term forecasting clearly is not an exact science!

When embarking on diversification there is some risk, but the position is stabilised by the company's core business. The study by Constable (1986) showed that in UK diversification has been accompanied by concentration through acquisitions and mergers to a greater extent than in other countries, and he makes two important points:

- the illusion is created in the 'league tables' for the large Groups of growth, much in excess of real growth
- investment in acquisitions is diverted from potential internal development which would produce real growth

1.7.5 Revolutionary change

The strategic planners have to be vigilant for first signs of revolutionary change or 'step functions', arising from political or technological causes. It is said that only one of the major oil companies had developed, in their series of future scenarios, a 'worst case' that forsaw the OPEC action of the early 1970s. But when this event happened, they already had strategic plans thought-through and pigeon-holed, ready to use.

Bergen (1984) has discussed how the 'catastrophe theory' model can deal with inventive discontinuity. The 'futurology' approach of the late Herman Kahn and others, and the lateral thinking exercises of Edward de Bono, are useful material to help company executives to break out of their customary 'tunnel thinking'. This is a general national problem amongst the mature countries of Europe, which tend to feed upon their cultural inheritance and past greatness. It is also a tendency in some of the technology Institutions, some of which are well into their second century.

The mature company, also, is at a disadvantage, as it suffers from the inertia of its inheritance of obsolescent resources, such as buildings that are less suitable and flexible than modern accommodation, together with a collection of barely adequate and aging plant and technology, and a long series of older products that have to be supported with spares, and occupy manufacturing space with obsolescent tooling, stocks and paperwork.

It takes courage to withdraw from a market sector early rather than late, but to do so can avoid the confusion of later collapse and heavy loses.

While the pursuit of profit can be carried to excess, it is a measure of the worthwhileness of an activity. Ansoff (1968) quotes what was said in the 1920s by Alfred P. Sloan of General Motors:

> . . . The strategic aim of a business is to earn a return on capital, and if in any particular case the return in the long run is not satisfactory, then the deficiency should be corrected or the activity abandoned for a more favourable one

Markets associated with information processing are especially volatile. For example, the good news for users of integrated circuits has been the availability of a continuing improvement in capability (Figs. 1.9 and 1.12), but for the manufacturers of the devices, the bad news was that they faced a rapid fall in the world market prices. Similarly with computer-aided design systems, the availability of 32-bit PCs and the process of windowing makes it less necessary to have expensive specialised terminals, so that the business of those who supply them collapsed.

Network systems for data reached the point where they became economically feasible, so altering the potential pattern of communication and detailed organisation structure within large companies. This development had also been held

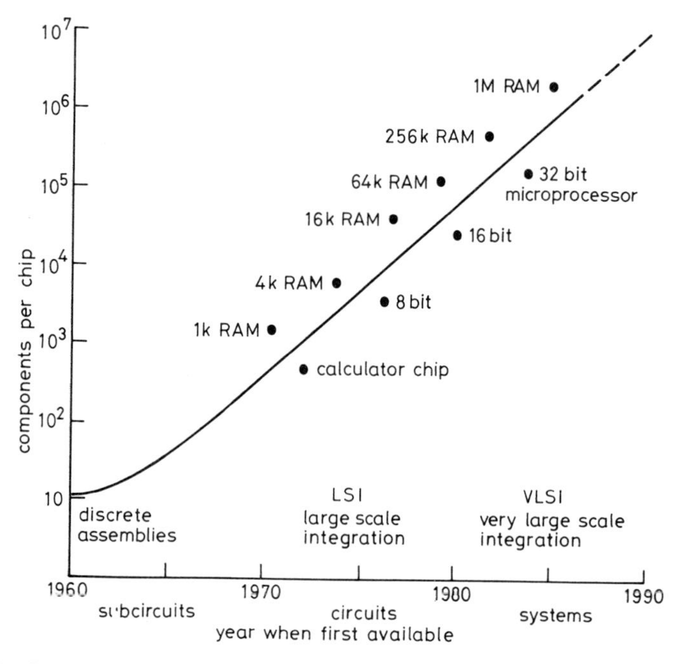

Fig. 1.9 ***Continuing improvement of the capabilities of integrated circuits*** **(*from Siemens*)**

back by lack of standardisation, but this can now be solved cheaply with low-cost memory devices providing dynamic interfacing.

Changes can be forseen, as Remhardt (1984) of Volvo has described. The benefit from the use of robots in manufacturing follows from the implications for at least three factors:

- reduced labour content of operations, and its social implications
- consistent high quality, a great reduction in random defects
- flexibility, because 'retooling' for product variants is via software, without physical hardware re-equipment

Although these technologies are new, revolutionary change is as old as history, and has been traced in technology through the last century by Freeman (1985) who reproduces the chart in Fig. 1.10 for sources of energy in industry, and Twiss (1982) has reviewed the socioeconomic influences of microelectronics.

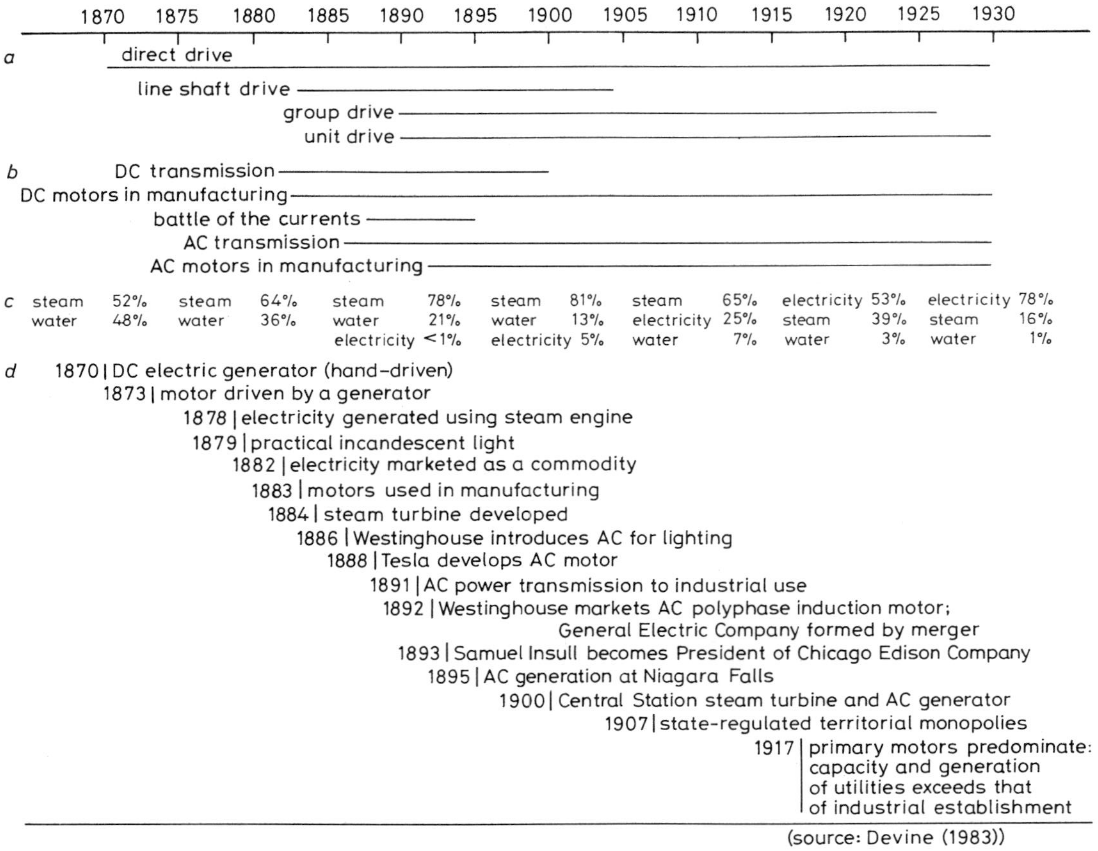

1870 1875 1880 1885 1890 1895 1900 1905 1910 1915 1920 1925 1930

a
direct drive
line shaft drive
group drive
unit drive

b
DC transmission
DC motors in manufacturing
battle of the currents
AC transmission
AC motors in manufacturing

c
steam 52%, water 48%
steam 64%, water 36%
steam 78%, water 21%, electricity <1%
steam 81%, water 13%, electricity 5%
steam 65%, electricity 25%, water 7%
electricity 53%, steam 39%, water 3%
electricity 78%, steam 16%, water 1%

d
1870 | DC electric generator (hand-driven)
1873 | motor driven by a generator
1878 | electricity generated using steam engine
1879 | practical incandescent light
1882 | electricity marketed as a commodity
1883 | motors used in manufacturing
1884 | steam turbine developed
1886 | Westinghouse introduces AC for lighting
1888 | Tesla develops AC motor
1891 | AC power transmission to industrial use
1892 | Westinghouse markets AC polyphase induction motor; General Electric Company formed by merger
1893 | Samuel Insull becomes President of Chicago Edison Company
1895 | AC generation at Niagara Falls
1900 | Central Station steam turbine and AC generator
1907 | state-regulated territorial monopolies
1917 | primary motors predominate: capacity and generation of utilities exceeds that of industrial establishment

(source: Devine (1983))

Fig. 1.10 *Chronology of the electrification of industry, from 1870 (from Freeman, 1985)*

Discoveries and technical advances are an irreversible process, familiar to the scientist and engineer, but feared and not understood by the lay public. Baker (1985) of CEGB said that the engineers of the future will need a far sharper awareness of the political and social dimensions of their jobs: more time needs to be given to their post qualification training, and to early exposure to business management. Engineering alone is not enough for engineers: they need to see themselves in the broader context of society as a whole.

One recent 'step function' has been the remarkable response to the theme of 'excellence', initiated in the book by Peters and Waterman (1982), and stemming from the recognition that the international success of the Japanese has been by taking a quantum step forward in product quality.

1.8 Implementing and reviewing objectives

1.8.1 Evolution of 'MbO'

Most engineers and managers have had some contact with the practice of 'management by objectives' (MbO) in one form or another. As mentioned in Section 1.6.1 (Fig. 1.7), there is a close analogy with H.S. Black's (1934) rationalisation of the negative-feedback process of stabilisation. If a signal is received or an action initiated, and a sample of the result is fed back, then, if it is not a replica of the original command, the difference is applied to execute correction.

To extend the analogy, if the information fed back is used incorrectly it can aggravate rather than rectify the situation. In the electrical model it can cause instability and an oscillation of increasing amplitude. In the human situation, counselling by the superior can help, but, if handled badly, can be disruptive.

Reddin (1971) defines MbO as:

> The establishment of effectiveness areas and effectiveness standards for managerial positions and the periodic conversion of these into measurable time bounded objectives linked vertically and horizontally with future planning, and the corporate plan.

MbO is a vehicle for carrying the corporate plan down through the management hierarchy, level by level, expanding the grand design of the corporate plan into the detailed tasks that will implement it. Ideally the downward communication should be complemented by an upwards flow of inputs into the next cycle of planning. The hierarchy of objectives must relate directly to the hierarchy of functions in the organisation structure. It is unwise to attempt to apply MbO in isolation in only one segment of the whole.

The main contributors to the concept of MbO were Peter Drucker (1955), John Humble (1965), Odiorne (1965 and 1979) and Reddin (1971). Their practices developed in parallel, mainly in the UK and USA, so there are significant variants:

- John Humble's: integrating the company's and the manager's needs
- George Odiorne's: common goals of superior and subordinate

- others have seen MbO as providing a 'contract' between superior and his subordinate, either in a cooperative vein or as a hard discipline
- also, primarily as a downward extension of corporate planning, and its implementations
- also, by the personnel function, as primarily a method of appraisal and development
- additionally, 'group MbO' has been put forward as a means of gaining the involvement and commitment of a group in formulating and implementing a course of action

Reddin says that in theory MbO looks like the only way to manage, but in practice it often can become a bureaucratic overlay, and a disguise for autocracy.

1.8.2 Drucker's contribution

For a clear understanding of the potential power of MbO, it is helpful to be familiar with both Drucker's approach and his background. He is said to have been the most influential and prolific writer on management (over 15 books), but he has remained outside the world of the Business Schools, while having extensive business contacts as a consultant rather than as a manager. He made an internal study of General Motors in Sloan's time, and later advised General Electric, but was never a businessman.

Recognising this, he titled his autobiography (1978) 'Adventures of a Bystander'. He is a very readable story teller, retailing parables and aphorisms to make his points. His facility with ideas stems from his multicultural background: born and bred in Austria, then as a young man working in London in the early 1930s with a traditional firm of merchant bankers: then to the USA before the Second World War, at the time of Roosevelt's New Deal recovery plan, initially as a correspondent for some UK newspapers, and as an adviser to European financial institutions.

On the strength of his first book 'The end of economic man' he was offered a teaching appointment at Bennington College in 1942, and his second book 'The future of industrial man' led to an invitation in 1943 to make a study of General Motors policies and top management structure. This was a part-time project over two years, leading to his 1946 book 'Concept of the Corporation'.

This appeared at a time when very few 'management' books had been published. Drucker began to fill this gap with a series of 'best sellers', in particular 'The Practice of Management' written in 1954, where in his preface he said:

> We have available today the knowledge and experience needed for the successful practice of management. But there is probably no field of human endeavour where the always tremendous gap between the knowledge and performance of the leaders, and the knowledge and performance of the average, is wider or more intractable... (the) first aim is to narrow the gap between what can be done and what is being done...

He broke new ground with his chapter 11, 'Management by objectives and self control', where he drew on his own wide cultural experience and business insights.

We learn from his biography how his teachers in fourth grade in Vienna early this century provided the basis for what became the practice of MbO, 50 years later.

Miss Elsa would set out a programme of goals each week with a pupil, and they would agree on action, note it down and each keep a copy, reviewing the programme the following week, and agreeing together the next week's programme. They kept a work book, made work plans and performance sheets, and set standards. The emphasis was on identifying an individual's strengths and building on them, in their own style: her pupils practised self-control and developed themselves.

1.8.3 How MbO is used

Personnel Departments in business and industry operate systems of performance appraisal, and some have seized upon MbO for this purpose, while others have adapted it as a tool of training and management development. In the UK, the Fulton Report (1968) on the Civil Service recommended organisation where possible as 'accountable units', and that those engaged on administrative work of this kind should know what their objectives are. It stated that the principle to be applied here is management by objective. In some branches of the public service there was initial rejection by Staff Associations and Unions, as they were uneasy how the criteria for assessing performance might be used, or misused. A later development for assessing effectiveness has been 'value for money', audits, and unit-cost comparisons between similar operations at different locations.

The outline in Chapter 2 (2.4), 'Review, appraisal and coaching' describes what is generally regarded now as good business practice in the use of MbO, with particular reference to the 'people' aspect of the person/boss review and counselling aspect.

In Section 3.7.1, Fig. 3.4 illustrates how MbO works in the 'corporate' sense, providing the mechanism for upwards and downwards communication in the hierarchy of the organisational structure of an operational management.

With project management (Section 1.10) there is usually an open situation where all members of the team have access to the information on goals and performances; so there is peer pressure to achieve the agreed levels of performance.

The personal review procedure should also continue in parallel, through both the 'project' and 'discipline' channels, so that an individual is personally counselled by the appropriate 'boss'. Clarity is needed here, so that a person may be clear what is his relationship, with which superior.

The main distinction between the 'corporate' and the 'people' aspects of review is in frequency. The 'corporate' progress review will probably be monthly or, in project situations, weekly. The major 'people' reviews are usually on an annual basis.

When MbO has failed in the past, it has been through inadequate preparation or over-elaboration – documentation extending to 20 pages has been known. The practice recommended by John Humble (1970) is to focus on the few key areas of results; those aspects that are the most critical at the time, and will have the greatest impact. Everything else can be left to the commonsense of the

individual, and covered in standing instructions, procedures and
reference of the job. A reassessment has been made recently by Seyna (1986).

1.9 Managing change

1.9.1 On being overtaken by change

Engineers spearhead change by technical innovation and, as managers, must establish a balance between cause and effect. Their technological planning must be accompanied by organisation, social and environmental planning.

Historically, there are many examples of failure to foresee the consequences of innovation. On 15th September 1830, the ceremonial opening of the Liverpool & Manchester Railway was marred when one of the distinguished guests, the local MP, William Huskisson, died after being run over by the Rocket locomotive.

The essence of safety practices is being aware of the dangers, and, from this and many other unfortunate incidents, the official Railway Regulations were built up. The sources were mainly from the conclusions of Inquiries into incidents, rather than from foresight. The IEE 'Wiring Rules and Regulations' were the result of foresight. After some difficulties with fires, and one of the first electrical fatalities at Hatfield House in 1881, where a pioneer installation of 117 arc lamps was operated in series and unearthed from a 5000 V generator, a Committee was appointed on 11th May 1982 and reported six weeks later, as described by Bowers (1985). The publication of the 'Regulations' avoided the situation in the USA, where in the years 1893–84 some 500 fires were traced to electrical causes, and there was great concern among insurance underwriters.

The first revision of the IEE Wiring Regulations was published in 1897, the 10th in 1934, and the 12th in 1950 introduced the fused 13 A plug and domestic ring-main system. The 14th Edition appeared in 1980.

It is instructive to review the causes of disasters, such as Flixborough and Chernobyl. Air pilots are avid readers of the reports of the Chief Inspector of Accidents, published by the Civil Aviation Authority. Quite mundane and previously unforseen events could have serious consequences. It was reported in *Pilot* (May 1986) that a flight engineer of a Boeing 747 (just one of a world fleet of some 700) had a mug of tea on his desk, and it was overturned: the tea drained into the lower circuit-breaker panel, and arcing was observed, and the circuit breaker of one of the two main generators tripped: a modification has now been raised to provide a cup holder on this table in all B747s.

Much that was first discovered pragmatically can now be established by analytical design. For example, we have ways of avoiding structural failures such as occurred with the Tay Bridge, the wind-induced oscillation of the Tacoma Narrows Bridge, and the fatigue failures of the early Comet aircraft.

The fiction writers demonstrated that quite accurate scenarios could be prepared: for example, H.G. Wells in 'Things to Come', Nevil Shute in 'No Highway', and Arthur C. Clarke's prevision in 1945 of broadcasting from orbiting satellites, which the Goonhilly Downs team put into effect in 1962 with Telstar and the first transatlantic television transmissions.

People are overtaken by social and environmental change, as when job skills become obsolete, or there are climatic changes, as in Africa. These aspects of change need to be tackled at governmental or international level. The events are fairly predictable, and can be alleviated without high technology by forward planning.

1.9.2 Stimulating change

The general euphoria surrounding new technology has both a good and bad influence. Public imagination and support are aroused, but if the follow-through is neglected, a counter-reaction may develop.

Britain had the first high-definition television service in 1936, in part through excellent project development by Marconi–EMI, but also the complementary contribution of John Logie Baird whose flair for publicity had captured the head-lines from the early 1920s. Consequently a 'market pull' was created, and the Government was conditioned to encourage the innovation, although it is said that the engineer-architect of the BBC, Sir John Reith, was not enthusiastic.

The technical role of Baird has been debated in the columns of the *Journal of the Royal Television Society* over recent years, but there is no doubt that he was a stimulant. Sir Clive Sinclair has occupied a similar role in relation to home computers, where the domestic penetration in the UK is the highest in the world, and this and the BBC Acorn have conditioned the next generation of young people to be computer-literate. Unfortunately neither Baird, Sinclair nor Acorn were able to achieve lasting commercial stability, as originally established.

1.9.3 The need for risk analysis

It is not easy in a fast-moving technical situation to ensure that all factors are kept under control. Where public health and the environment are at stake 'rules and regulations' are likely to acquire statutory backing, as the consequence of such incidents as:

- 1912: Titanic, 1403 lost
- 1934: Gresford Colliery, 264 deaths
- 1930: R101 airship, 47 casualties
- 1937: Hindenburg airship, 36 deaths
- 1966: Aberfan earth slip, 147 people, mostly children
- 1970: Yarra Bridge collapse, Melbourne, Australia, 35 deaths
- 1974: Flixborough Nypro plant, 29 deaths
- 1979: Sevaspol dioxin plant
- 1983: Three Mile Island, Pennsylvania: no deaths
- 1984: Bhopal pesticide plant: 2000 deaths
- 1986: Chernobyl, Russia: over 30 deaths directly attributed
- 1987: Herald of Free Enterprise ferry capsize, 188 deaths
- 1987: Kings Cross escalator fire, 30 deaths
- 1988: Piper Alpha North Sea platform, 169 deaths

These are all engineering-related incidents, to which could be added a number of serious dam disasters. Many disasters attributed to natural causes could have been foreseen and mitigated by the application of scientific knowledge. It will be noted that the direct casualties in the energy-producing industries were considerably higher when coal was the principal source.

Risk analysis as a technique has become highly developed, and is a key element in managing change. Additional costs are incurred, but these have to be seen as an insurance premium against the consequences of potential disasters. There are no specifically acceptable levels for risks, and their avoidance costs have ultimately to be borne by the consumer.

Costs to the community also arise when an industry is being run down. It is reasonable to expect the Government to provide subsidies to ease the transition period, but then it is unrealistic to maintain them long term against world trends. The cost to the community of phasing out an obsolescent industry can be minimised by forseeing the need and taking early action, despite the political difficulties.

1.9.4 Organising for change

An organisation contemplating a major change or innovation has three main tasks:

- perfect the particular technology
- plan for its introduction
  - capital resources
  - physical facilities
  - people
- Consider its secondary effects on the community and the environment

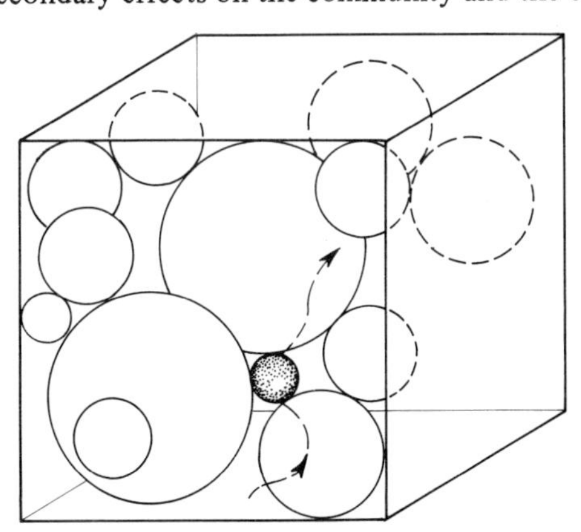

Fig. 1.11 *The industrial economy can be represented by this model, the balls each being proportional to the size of a particular company. The large companies are virtually fixed in the economy, but the smaller ones can migrate easily, particularly the very small ones. As companies expand, they become less mobile, unless the economy also expands (from Johnston, 1957)*

The scale of modern technology is such that major developments have to be thought through on a multinational or global basis at high cost.

But in the presence of the big ventures, there are opportunities for smaller enterprises that are fast moving and mobile, to fill the gaps that appear. The simple model in Fig. 1.11 illustrates the mathematical problem of securing the greatest 'packing density' with optimum proportions of large and small enterprises, shown as spheres within the cube that represents the total national product (Johnston, 1957). The scope for the smaller spheres is considerable, when we remember that a single large sphere leaves unoccupied nearly 48% of the volume of the cube.

Engineers have to adapt to greater rates of change than most other professional groups, and this is particularly so in electronics and computer systems, and in the development of integrated circuits and memory devices. As mentioned in Chapter 5 (5.4), 'intercept techniques' are applied to forecast what will be the cost of key elements of a product, by the time of market release. The trend is reasonable continuous in the example of Fig. 1.12*a*, but there are discontinuities if a new device becomes available, as in Fig. 1.12*b*, within the design life of the product, which may then suffer early obsolescence.

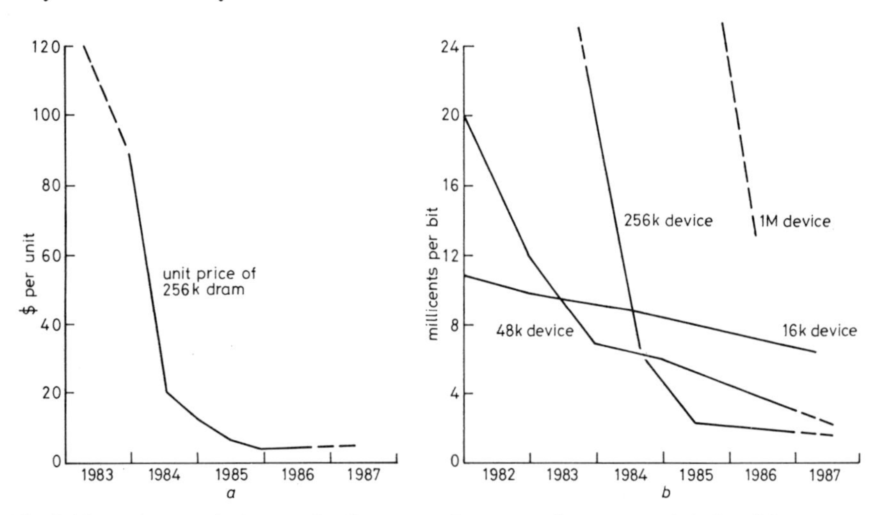

Fig. 1.12 a *A new device can be 'intercepted' at an early stage, and designed into a new equipment in anticipation of the cost reduction curve, characteristic of semiconductor devices*
b *The availability of new devices quickly makes obsolescent equipments introduced only some two years earlier: for mass memory, devices of lower capacity remain competitive, but are larger and more costly to assemble*

A similar problem arises in constructing major plant, where the time-scale may be 5–7 years from first design: on completion the control systems built into it may already be obsolete technology. This is no new problem, as it happened in aviation with the Brabazon and Princess flying boats, when the full development of adequate engines was not realised in the same time scale as the airframe, and again with the Nimrod where the airframe was ready, but not the radar-processing system.

1.9.5 Discerning trends

After 25 years of editing the *IEEE Transactions on Engineering Management*, A.H. Rubenstein (1985) looked ahead to identify trends in R & D and in engineering innovation, with these conclusions:

- technical entrepreneurship: a high-risk game, and we shall have to learn how to foster it
- corporate research laboratories: there is a trend away from central units in large companies, towards decentralisation and applied work, giving fewer opportunities for advanced research
- divisional networks in large companies: these can compensate for lack of a central corporate function, and need to be linked to better long range planning, and definitions of corporate strategy
- evaluation of R & D results: if R & D managers do not do this for themselves, it will be done for them, to their disadvantage
- make or buy decisions: there is a trend towards drawing on outside resources, by obtaining licences, and using technical consultants and software contractors – the danger is that this creates gaps in the company's capabilities
- background of the Chief Technical Executive: increasingly coming from other functions, such as production, marketing or general management

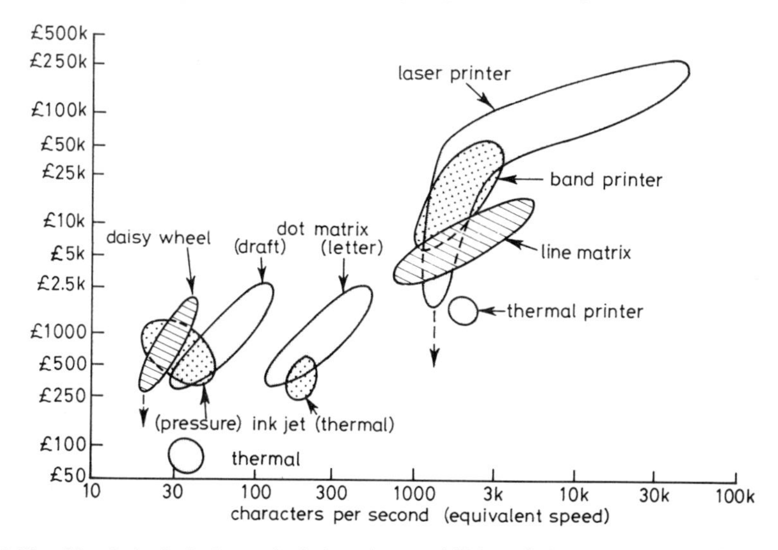

Fig. 1.13 ***Morphological chart, depicting the capabilities of the several types of computer printer, against price***
It is a changing relationship with time, as the laser types of printer are becoming cheaper, while the other types are fairly stable in price

On a wider canvas, Drucker (1971) provided an 'early warning' that the relative stability of the last century was being overturned by four discontinuities that will shape the future:

- new technologies, creating new industries and business opportunities
- major changes in the world's economy, and balance of the trade between nations
- social tasks are being entrusted to institutions run by managers, which many younger people reject
- knowledge is now the crucial resource of the economy – power and responsibility now lies with those who constitute the knowledge society

In his book 'Megatrends', Naisbitt (1983) highlights in popular style ten new current factors for change.

At the detailed level it is helpful, at least to engineers, to depict relative changes with graphics, e.g. in Fig. 1.13, which shows a snapshot of the relationship between price and performance of classes of printer. If a series of annual plots are animated in sequence, the relative rates of change stand out: for example, laser printers are expanding their scope and becoming cheaper. This technique, known as morphology (borrowed from classification in biology and philology) has been described by Shurig (1984) of Ontario Hydro.

1.10 The Project concept

1.10.1 Projects – vehicles of change

It is more comfortable to assume that there will be change, than that there will not. Change brings uncertainty, but this can be minimised by specifying a limited objective or milestone and concentrating on achieving it. Having climbed one hill, we can then more clearly see the next.

There is a strong case for managing nearly all activity according to the project concept, using each completion target as the stepping stone to the next. This analogy applies as much to personal career planning Chapter 2 (2.6) as in business ventures. The examples of new ventures in Chapter 6 illustrate how a business can start off on the basis of one set of assumptions about its place in the market, and then adjust to new objectives as the potential for them is identified.

In this part of the book we address the main concepts underlying project management, but without attempting to detail the management techniques employed, which are explained in companion volumes in this series, by Twiss (1987) and by Hussey (1987).

General understanding of project-management principles is bedevilled by the variety and range of magnitude of activities that are termed 'projects' – to quote the Concise Oxford Dictionary:

> *project*, n. Plan, scheme; planned undertaking, especially by student(s) for presentation of results in a specified time

A good student project may occupy 100 hours of work. A major power station design and construction project may require more than 10 000 man-years (20 million man-hours) on site plus nearly as much effort off-site in design, manufacture and supplying services, at an overall cost of up to £1000 million. The cost of introducing a new major aircraft is similar. The original Apollo moon programme cost $20 000 million (in 1960s dollars), and involved 5000 industrial companies and some 300 000 technical staff.

An account of the 25 years of NASA (National Aeronautical & Space Administration) by Beggs (1984), until recently the Administrator, quotes the first Director, John Webb:

> In the large-scale endeavour the man himself must also be unusual: he must

> be knowledgeable in sound management doctrine and practice, but able to to a job without an exact definition of what it is or how it should be done; a man who can work effectively when the lines of command crisis-cross and move in several directions rather than straight up and down: one who can adjust to, and be himself, several bosses at the same time; one who can work effectively in an unstable environment and can live with uncertainty and a high degree of personal insecurity; one willing to work for less of a monetary reward than he could insist on elsewhere; one who can blend public and private interests in organized participation to the benefit of both.

As a project-based undertaking, NASA enjoyed continuing success: some recent difficulties in the Shuttle programme seem to have been associated with the change to a more routine operational role, linked with deteriorations in internal communications.

During the 1960s and 1970s, the UK had periods of difficulty with large construction projects, due to the interplay of communication problems and uncertain industrial relations. In his review of the Grain project, Burbridge (1984) shows the way towards structuring large projects so that tasks and roles are sufficiently clearly defined that success does not entirely depend on the availability of a 'superman' as supremo. The characteristics of leadership Chapter 2 (2.7) can be in the dominant style used by the late Admiral Hyman Rickover in the original nuclear-propulsion submarine programme of the US Navy, or it can be by concensus, and coaching and guiding subordinates to become a self-motivating team.

Major projects can last 5–7 years or more; so no one is likely to have the experience of serving in a senior position in more than two or three, in their whole career. The project style of management therefore must be taught in order to accelerate the learning process, and among the first such programmes is the Fellowships in Systems Management, initiated in 1985 by The City University, in conjunction with the EITB.

The term 'systems management' has recently assumed a broad meaning, as defined at a recent Colloquium (1986) arranged by the Project Management Forum, an association of five of the major engineering Institutions. It is seen as a comprehensive discipline embracing within its compass strategies, planning, design engineering and project management.

This wider context is not to be confused with the purely technological and analytic practice of systems analysis and design.

1.10.2 The organisational basis for projects

As we have seen, there is a tremendous range in the size, duration and complexity of projects. A linked concept is that of 'matrix management'. The concept has been known for some 25 years, and is reviewed in a book by Knight (1982). It is said to be rarely applied, and to be heartily disliked by many managers because of its complexity: a positive way to look at the matrix concept is as a model that may be approximated to when conditions are favourable.

For example, in an R & D organisation the model would be as in Fig. 1.14. A number of 'discipline teams' may be the base for internationally respected authorities in particular specialisms. For each of these vertical axes there is a Head of Department. From this resource staff and/or services are provided to a number of current projects, shown as the horizontal axes, each with a Project Manager. Two Directors are shown. One controlling the Departments, and the other the Projects. For this to work well there must be high standards of professional conduct, and a minimum of 'internal politics'. Particularly damaging to the concept of the model is any suggestion of a power struggle between the two Directors.

director of projects — — — — — — — -director of research

project managers | department heads

	vacuum physics	metallurgy	control systems	chemical engg.	workshops	administration
project A	$\frac{1}{2}$	xx	x	xx	xxx	xx
project B		x	xx	xxx	x$\frac{1}{2}$	$\frac{1}{2}$
project C	x$\frac{1}{2}$	$\frac{1}{2}$	$\frac{1}{2}$	x	$\frac{1}{2}$	$\frac{1}{2}$
project D		xx	x		xxx	xx
minor projects group	xx	x$\frac{1}{2}$	xxx$\frac{1}{2}$	xx$\frac{1}{2}$	xx	xx

Fig. 1.14 *Model of matrix management in an industrial Research Division, showing (x) staffing requirements of projects, as a basis for human-resource allocation*

The advantages are that depth of expertise can be made available, experts can be phased in and out of projects as they are required, but each project has its own team identity as a multi-disciplinary group appropriate to each project. The younger members can gain much in experience by participating in a series of teams. This is the ideal situation, with a condition of balance between the two axes of activity.

Although the concept is most attractive, in practice the operating difficulties include:

- initially, introducing the concept, when people find it unfamiliar and may reject it
- the idea of being responsible to two bosses is only slowly accepted, as people adjust their game to the new rules
- the general tone is set by the two Directors (or their equivalent), either for better or worse
- resource allocation of both facilities and people needs to be properly planned, but it is complex, and can be upset by changing priorities
- project managers may be unsettled by conflicting priorities on the finite resources
- individuals may be unclear which of their bosses really influences their career

prospects, particularly if their time is spent on more than one project, and there is no system of annual review
- when a project lasts a long time, team members tend to lose their 'discipline' links and expertise
- if an individual has to serve several projects, there is a feeling of reduced loyalty to any one of them

The matrix idea therefore works most effectively when individuals can concentrate on only one or two projects, and where the turnover of projects is relatively rapid, so that they come as progressive assignments.

However, the concept is more soundly based if the rigid assumption of a permanent establishment is relaxed:

- an individual qualified in a particular discipline can plan his career path as comprising periods of project experiences, either by seeking internal transfer, or by selectively changing his employment
- a company can supplement its permanent staff by taking people in on short contracts or from Agencies, for the peak periods of particular projects.

This greater mobility is experienced in the USA, and it makes for a more dynamic atmosphere. It tends to release the interpersonal tensions that can build up where there is long-term job tenure. The software industry has led in the use of freelance and Agency staff, and this pattern may extend further into other fields of technical employment.

With greater mobility available to discipline specialists, their managers will need to give more attention to personnel management. Sometimes the head of a discipline is not too keen on the administrative responsibilities that go with his seniority, which is something that his own superior must try to correct.

Sheane (1977) quotes Robert Malpas, then a director of ICI and now Managing Director of BP:

> The matrix works better if there is a clear overall objective, not evolved from the 'power centres'. They do not work if behaviour is immature. A key management skill is the ability of an individual to play different roles, and the required roles must be defined explicitly. The management information system must mirror the matrix and present costs both by product and by territorial market. Federal structures will work if these needs are met:
> - better thought on strategy and on 'selling' objectives
> - clearer role definition, to secure better and more mature behaviour
> - better management accounting
>
> Compexity has to be managed, and the necessary skills can be taught

Knight quotes a diagrammatic presentation of influence in matrix organisation, originated by J.R. Galbraith, and shown in Fig. 1.15.

At one end of the scale, the 'function' has total authority, and at the other end the 'project manager' is in full charge. Actual situations will be somewhere between the two. Taking Malpas's advice it is better for the roles to be defined than for the balance of influence to be determined by the force of personality of individuals.

The project leader is most effective when he has full authority as a manager, as then he has a virtually full-time team. The opposite situation often occurs in part-time projects, where the project leader's role may be no more than that of co-ordinator, administrator or facilitator. Even so, a persuasive, diplomatic individual can be very effective.

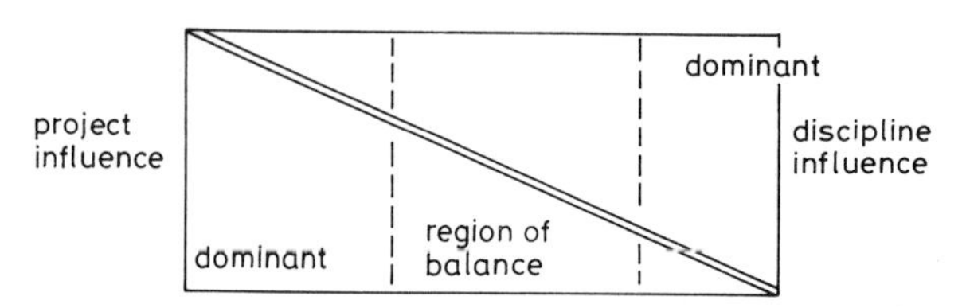

Fig. 1.15 *Relative influence of the 'project' and 'discipline' axes of the organisation matrix of Fig. 1.14*

Wearne (1970) explained the 'influence' parameter as a 'balance of authority': this he represents as a vector diagram, which will appeal to electrical engineers as a model of 'real' and 'imaginary' power, related by phase angle.

Malpas's point was that, whatever the balance is, it should correspond to the policy intentions and not be the result of immature manoeuvres among staff.

In what is now a classic study, Burns and Stalker (1961) looked at a number of companies in the electronics industry during the 1950s. They found that the successful ones were those that were very loosely structured, with overlapping roles, with emphasis on horizontal rather than vertical communication, and authority based on knowledge rather than seniority. They termed this style or organisation 'organic', being the opposite end of the scale from 'mechanistic' – the conventional (or 'military') hierarchical command structure.

The negative view of the 'organic' philosophy is that it is not possible to run a large organisation this way. But positive thinking leads to the solution that it can become a federation of smaller, more highly motivated units, contained within the matrix framework.

The sociological implications of this model appeal to many people, as providing an escape from 'establishment conventions', and providing opportunities for freedom and self-expression. A way has to be found to relate these aspirations to the commercial need to be results-oriented.

In recent American experience of companies in the semiconductor and computer-related fields, combined models of organic/mechanistic styles have been called 'loose/tight'. This means that overall corporate goals and control systems are in place, but within the terms of reference and budget of a particular mission there is considerable freedom.

Similar thinking underlies the pioneer work of Volvo in Sweden from the early 1970s, replacing the car assembly line with assembly teams who are self-organised. Productivity rates are about the same as before, but quality is enhanced, and it has been easier to attract and retain good workpeople. The emphasis on self-determination in the 'organic' model is probably the key difference between the oriental and Western attitudes to work.

The book by Lock (1977) is a British survey of project management, and those by Cleland (1983), Kerzner (1981) and Stuckenbruck (1981) are representative of comprehensive American texts. There is an International Project Management Association (INTERNET), and its *International Journal of Project Management* is published quarterly by Butterworth Scientific.

1.10.3 Team building

There is a relative permanence about attachment to a discipline department, and the internal relationships will be of a long-term nature.

A project leader or manager is in the opposite position. There is a need very quickly, at the start of a project, to knit together the people available to him, as an effective team, making the best use of their diverse prior experience and skills. He cannot expect of everyone available that they will be the ideal candidates for the roles to be filled.

Taking as a typical model, a medium-sized project with a programme likely to run for 1–2 years, the team available might come from half a dozen disciplines, each contributing a full-time member plus others part-time.

Assuming that most of them have not worked together before, then the project manager has these initial tasks:

- introduce the team members to each other
- outline the broad requirements and programme for the project
- determine 'housekeeping' needs, and the requirements for other services
- decide who does what

A complication is that complete teams do not usually start together on day 1. There may have to be two or three progressive versions of this session.

Rudyard Kipling struck the right note when he wrote:

> I keep six honest serving men
> they taught me all I knew:
> Their names are What and Why and When
> and How and Where and Who

Unlike the head of a discipline, the project manager is not the 'chief expert': nor need he be the most senior person from the point of view of staff grades. His expertise lies in drawing upon the skills of his team, and helping them to solve the problems. He is a 'facilitator' rather than a superior. In the initial stages the 'organic' approach can be powerful, and he can use it to:

- bring the collective experience of the team to bear on the detailed needs of the programme
- discover who has relevant prior experience that can be helpful
- resolve by discussion and agreement who does what, particularly where responsibilities may seem to overlap

- identify areas of 'underlap', where needs are not met yet, and resources are inadequate
- clarify objectives in everyone's minds
- build up motivation and team spirit

This process can be conducted in informal sessions, and it is helpful to start by building up the agreed structure of the project on a large matrix chart (Fig. 1.16).

	key area of results ↓ / function →	general manager	plant	mining	central engineering
output	developing natural resources 1	R		C	
output	operations— planning 2	R	C	C	
output	operations 3	A	R	R	C
output	controls 4	A	ROLE		
input	human resources 5	S	C	C	C
input	physical resources and services 6	A			R
input	policy and standards 7	S			A

Fig. 1.16 *Determining 'who does what' on a major project (in this example, a new metallurgical and mining development)*
For each contributor to the matrix, the 'role' is agreed: R, C, S, or A, as defined in the text: then any overlaps or underlaps are mutually resolved

There will be apparent gaps at first, signalling 'underlap', and the overlaps can be resolved by agreement on how the work load can be balanced through the team.

When more than one person seems to share a key area of responsibility, it is helpful to disect and identify complementary roles, for example:

R: responsible for results: takes initiative on action plans
C: contributes or makes use of this key activity
S: sets standards/monitors/audits
A: advises, is consulted, provides data

Initially the chart is worked up in outline with marker pens, and then a more detailed definitive version is published.

This approach can be presented as 'making your own organisation structure', and is a new experience to many participants, who will usually react with enthusiasm. Starting thus tends to discourage any tendencies to political manoeuvres, because what is agreed has involved all participants.

The book 'Corporate Man' by Antony Jay (1971) will be helpful to someone preparing to initiate a new project. Jay writes from his experience of bringing together teams of talented people on the programmes side of the BBC. There is a measure of 'show biz' in all spheres of management and leadership, and a good project manager resembles the impresario who can caste the roles, and get the show on the road despite temperamental behaviour by some of his star performers.

Jay also draws the parallel with a tribal hunting band – the project situation where the work is associated with tough construction-site conditions or extremes of climate. An insight into this kind of field engineering work is given in Auletta's (1984) book on the Schlumberger Company, where he describes the work, training and motivation of the field-prospecting instrumentation engineers.

After achieving a good start-up, a project manager must be alert to maintain the high motivation, against the attrition of day-to-day problems. The key to this is to have regular review meetings where these problems are identified, fed back and then resolved at an early stage, before they build up. The project manager and his immediate assistants will also make their own cross checks by the simple sampling technique of 'managing by walking around'.

One of the earliest and most successful project-team programmes, in the contemporary style, was launched over 50 years ago, and is described by Burns (1986). It culminated in the launch of the British 405-line television system in 1936 by the Marconi–EMI Television Co Ltd. The system was described in a paper presented to the IEE by Brown *et al.* (1938).

In the early 1930s the EMI company had assembled a strong team of engineers and scientists to develop the studio equipment, led by Isaac Shoenberg who previously had been head of the Patents Department at Marconi's. By June 1934 the team comprised 33 graduate engineers and scientists, 28 qualified technicians and laboratory assistants plus support services, a total of 108 personnel.

They developed electronic scanning in stages from a definition of 120 lines to 240 lines and then to 405, a system that was not finally withdrawn until 1980. Burns compares this successful approach with that of the competitive team at the Baird Television Company, who were late in changing from mechanical scanning to the vacuum-physics approach. Comparable teams existed in the USA, in particular at RCA, CBC and Bell Telephone Laboratories, but the public launch of the RCA service came three years later than that in the UK.

The History of Radar (1985) also contains a number of examples of *ad hoc* 'organic' and seemingly amateur teams, brought together at the time of the Second World War, often from a background of academic research, and including key people from the pre-War TV development teams, such as the systems and circuit designer, A.D. Blumlein.

On the manufacturing side this rather British style of improvisation paid off, because key system modifications could be implemented overnight. Comparable German equipment was very finely engineered, some being tooled up with die castings, but with enhancements taking much longer to implement.

Today, the 'organic' and slightly disorganised style of R & D project management is still to be found, e.g. in the description by Kidder (1981) of the development of a new computer (MV/8000) at Data General Corporation.

1.10.4 Extension of the project concept

The advantages of flexibility and effective motivation encourage other applications of the project concept. For example, in an ongoing year-to-year situation, there can be a new operating plan and set of objectives annually, which can be referred to as Project 1987.

The process of 'zero base budgeting', described by Bragg (1982) of Cooper & Lybrand, has similar features: the whole need for, and purpose of, an activity is reassesed, starting again from zero base. It has been valuable as the means of planning the provision of services and other indirect overhead costs.

When a new activity or new facilities are required, experience has shown that it is usually simpler and better to plan afresh on a clean sheet of paper, and set up on a 'greenfield' site. A new manufacturing unit can be launched on 'project' lines, preferably followed by a clear hand-over 'milestone', when the long-term operational management assumes control, and the project team withdraws. This is similar to what happens in the construction and commissioning of a ship or a complex chemical plant, or a new power station.

Present-day industrial sites and building construction are so different from traditional town-centre industrial buildings that it can cost as much to adapt the old as to acquire the new. An old central property often has a high realisation value for redevelopment, which may pay for a new building elsewhere. On the human side, a fresh team comes together in fresh surroundings, and management can guide them towards new attitudes and methods.

The success of the Japanese in setting up European branches in as much due to the 'greenfield' factor, as to the difference in the industrial philosophy current in Japan. This is usually adapted considerably for the British scene. The one common factor seems to be that an agreement for a single Union to represent all workers is included in the early negotiations. Much care is given to selection of staff, and to their initiation and training.

Computers are an essential tool for project programme scheduling, and for time and cost control, for, as Benjamin Franklin wrote in 'Advice to a Young Tradesman' in 1748: '. . . remember, time is money. . .'.

Now that a wide range of project management software packages are available for use on microcomputers, they can be a convenient means of controlling work programmes in general. Where possible, a large project should be dissected into a hierarchy of sub-projects, with a limited number of common nodes, as this is easier to control and delegate.

The best combination is to use data processing for the detailed scheduling, plus traditional wall charts to convey an overall impression, and to display the interfacing between sub-projects. Graphical methods should not be underestimated, as they serve as a 'language' and an engineer's shorthand.

It is useful to maintain a project office or project control room, even though the team members may spend most of their time elsewhere. A good chart display provides an accessible update of information, and the room is a focal point for the team's activity. An 'open' attitude is to be encouraged so that problems are known immediately. Those concerned should not be criticised for bad news, but helped to overcome the problem. As Oscar Wilde wrote in 'Impressions of America':

the notice said:

Don't shoot the pianist
he is doing his best

Some of the more recent studies of groups include that by Holroyd (1983) of the University of Bradford, who has extended the matrix concept to social networks, seen as communication systems, alternative to and supplementing the traditional 'hierarchical power structure'. He calls such sub-networks 'guilds', embracing clubs, societies, institutions and spontaneous groups. He provides a comprehensive bibliography of work in the social sciences.

Belbin (1982) has made a psychologist's analysis of why management teams succeed or fail, the causes being much more to do with interpersonal relations than with the type of organisation structure.

1.10.5 Examples of Projects

The informal Project Management Forum is supported by the Institutions of Civil, Mechanical, Electrical and Chemical Engineers, and the National Economic Development Office (NEDO), as a vehicle for sharing experience on the conduct of major projects. A series of Colloquia have been hosted by the constituent Institutions, which have shared the task of publication.

In recent years, power stations have been amongst the largest and longest-running constructions initiated in the UK, and have experienced the greatest difficulties, organisationally and with industrial relations. The continuity of policy and direction provided by the programmes of the CEGB has resulted in more effective strategies and procedures, as reported by Burbridge (1984) in reference to the Grain station, and at the Colloquium (1986) on the Drax completion project. The latter was completed at a final target cost of £900 million in five years, as compared with nine years on the earlier project. These schemes typically involve the coordination of 100 contractors and a 1000 suppliers.

A lecture by Sir Alistair Frame (1986) outlined the methods used by RTZ Group to manage capital projects throughout the world, a number of these being in excess of £100 million each in value. It is of interest that the speaker was team leader for planning the earlier Channel Tunnel scheme, cancelled in 1966. The revised project is estimated to cost in the region of £6 billion over seven years.

The Thames Barrier, described by Horner (1987) was completed in 1984 after nine years, at a cost of £440 million. For the next Thames road crossing, to supplement the Dartford Tunnel M25 link, it has been interesting to see that the same order of costs were estimated for three quite different engineering solutions, a bored tunnel, submerged precast sections, or a suspension bridge, at about £200 million.

The history of project management in the defence procurement sphere, was traced by Vice Admiral Sir Lindsay Bryson (1982). There is a recent trend in defence towards placing management responsibility with a prime contractor in industry: again, a typical project duration is nine years.

The Alvey Directorate has developed considerable expertise in the structuring of a suite of projects, in an overall programme of advanced research and development, and has introduced widespread co-operation between companies at the 'pre-competition' level of application development.

For projects of more modest size, down to a few thousand pounds, a useful contribution has been made in the schemes of financial assistance provided by the Department of Trade & Industry, such as Microelectronics Applications (MAPCON), QAAS (now business and technical advisory services, BTAS) and the Design Advisory Service to Manufacturers (DAS).

As the public purse is involved, the Department exercises a firm discipline over the definition and stage progress of each project. A useful spin-off from this is the thorough project management training that the participants experience, to which they may not previously have been exposed.

Applicants are required to define the proposed project very clearly, including purpose and method, the major problems to be solved, an assessment of chances of success and the criteria, and details of the planning and control methods.

A detailed costing and time programme is required. Estimates are requested of the sales of the product, the market size and the principal competitors. If for internal use, the expected benefits and the prospects for selling the know-how via licences are required.

Progress payments are made against evidence of satisfactory technical progress. The form of both the proposal and the reports is assessed critically by representatives of the sponsoring Department. An example of such a sponsored project is described by Hudson (1986), for the development and implementation of a flexible manufacturing system in his company.

1.11 Managing operations

1.11.1 'Operations' versus 'Projects'

The objectives of project management and of operations management are virtually opposites:

- *projects* create innovation and change
- *operations* replicate and maintain services and facilities to fulfil market or client needs, to uniform standards

Common ground between the two exists, because an operations manager may wish to develop different or better services, and can carry out the implementation of these changes as an 'improvement project'.

Good operational management exists when these four essential and complementary tasks are carried out at the appropriate levels (as set out in a recent booklet from BIM (1985):

- Board objectives, strategy and policies
- Top management action

- Effective operational management
- The marketing function

Wild (1980) defines 'Operations Management' as: 'the management of systems for the provision of goods or services'.

The term contains, but is wider than, production or manufacturing management: operations in a factory are a system concentrated physically in one location, while field operations, such as distribution, installation or servicing, are systems physically distributed over considerable distances and working locations. A further definition is:

> an operating system is a configuration of resources combined for the provision of goods and services. . . they convert inputs (from all functions) in order to provide outputs which are required by a customer

A functional specialist in an organisation needs an appreciation of the whole 'operations system', in order to co-operate and contribute to it in an effective manner. It helps to have a knowledge of the specialised vocabulary used by his colleagues, in order to communicate competently.

1.11.2 Managing operations

It can happen that a specialist engineer or senior scientist can find themselves in charge of an Establishment – they will then wish to know a little about every aspect of the operations for which they will now be held responsible. They will manage by delegating the responsibility to the heads of each function of operations, but have the need to know enough to ask the right questions, and to understand the answers, together with the routine management reports.

In large organisations it was customary to experience barriers and departmentalisation between the several functions. Sometimes the heads of functions would require all communications to be routed through them personally, and would discourage informal lateral contacts with other departments. In such an atmosphere, it was often said that things were done through the grapevine, not via the formal organisation chart.

Such styles of management are not appropriate today, but are still to be found here and there, and are out of harmony with the newer manufacturing and communications technologies. A judicious balance is necessary between the 'mechanistic' style, and the freer 'organic' style – whatever is done informally does need to conform to the overall 'rules of the game', and not ignore the required practices and formal allocation of responsibilities, or an imbroglio will ensue, if not anarchy.

It is good management practice to head off a 'balance of power' situation, and any internal politics, based on the personal aspirations of the heads of functions: this has been stressed by Malpas (1982) whose career has been at senior levels of both ICI and BP. If the 'rules of the game' are clearly understood by all the players, there should not be too much for the umpire to do.

1.11.3 Operations as a 'system'

What has transformed the traditional situation in large organisations is the advent of three aspects of new technology:

- local area networks (LANS)
- automated manufacturing techniques (AMT)
- flexible manufacturing systems (FMS)

Information and knowledge is no longer the property of a particular department (although they may be responsible for its accuracy). This is in marked contrast to 'the old days' when a supervisor or a foreman would have a 'little black book', which was regarded as his personal property, and knowledge was the basis of his authority.

For the future, corporate operations will be 'systems', to which the functional specialists in departments contribute, and information will be available at a touch of a key, to those who need to know.

The difference can be summed up in the two diagrams of Figs. 1.3 and 1.4: functions will be linked by LANS, diffusing together. The facility for communication provided by networks and protocols of LANs internally, and TOP externally (see Section 1.4.6) provide a common language, and removes the distinctions in traditional organisation structures between solid and dotted lines, vertical and horizontal lines, and internal and external communication.

The following Sections introduce the several functions within the complete operational system. The trend recently has been to 'professionalise' every recognised function in business operations, so that a 'body of knowledge' is developed in each, as in the traditional learned professions of law, medicine, science and engineering.

1.11.4 Purchase and supply

This function is known as procurement in the USA, and 'materials management' where the emphasis is on handling and distributing bulk commodities. The UK's Institute of Purchasing & Supply originated in 1932, and has about 16 000 members, and there is a scheme of professional examinations. The business of matching needs to sources has become highly developed, changing from a style of confrontal bargaining to co-operative achievement of optimum value for money.

The technology of products and materials has become more complex, and sourcing locally is replaced by global supply, coupled with tighter control on quality and specification. What was merely processing of orders has become the management of strategic supply. Hadnam (1985), who is Purchasing Director of Perkins Engines, has published an account of the strategic approach. The 'just in time' concept greatly reduces stock levels, making use of direct communication links with suppliers, and manufacturing software systems remove clerical effort (and errors) from the processing of bills of material, purchase orders, stock control production scheduling and inventory control, and also provide the relevant cost information. A colloquium reviewing professional purchasing was conducted by Critchley et al. (1987).

Flexible manufacturing makes it possible to process customers orders through manufacture or by material conversion very quickly, as it is less necessary to make for stock in economic batch sizes. Statistical methods are important adjuncts to inventory control systems, and are described in Part B of Hussey's (1987) book. There is a BIM Checklist (no. 77) on the management of purchasing.

1.11.5 Production management

Production engineering and industrial engineering have a long history, going back to the pioneer work of Taylor, Gantt and Gilbreth before the turn of this century. The Institution of Production Engineers was founded in 1921. With the rapid development of computer applications in manufacturing, production engineers are much concerned with information technology systems. The main engineering Institutions now arrange joint events through the Engineering Manufacturing Forum. There are books on the quantitative aspects of production management by Bestwick and Lockyer (1982), Schroeder (1982) and Gaither (1984).

1.11.6 Physical distribution

This function, together with procurement, constitutes the total process of logistical supply, a term originated in the military world. The BIM supports a Centre for Physical Distribution; the function has become highly developed, and includes the techniques of automated mechanical handling. As manufacturing processes become concentrated on a few highly automated plants, the cost of packaging and distribution assumes a higher percentage of the total costs.

Mathematical modelling techniques, based on Operational Research, have been applied since the early 1940s, and are used, for example, to optimise the overall distribution systems, by identifying the most suitable principal and secondary nodes for a distribution network, given such constraints as the locations of motorways and ports, and the business hours of customers' warehouses.

The development of standard freight containers has transformed the international transport of everything except bulk commodities such as ore, bauxite and oil. The main costs are at the terminal interfaces, with distance a relatively minor factor in the cost equation.

Standard containers and pallets for air freighting have been developed in a similar way. Although air freight costs more, the standards of handling and protection are good, so that is attractive for shipping high-value items such as computers and microelectronics: the quick delivery allows stockholdings to be minimised. There is a BIM Checklist (no. 33) on Physical Distribution Management, and several others on aspects of these operations.

1.11.7 Field servicing

The majority of users of complex equipment do not possess the skills to carry out either preventative maintenance, or to deal with breakdowns. For computers remote diagnostic routines are available, but for most other equipment physical attention is necessary. A significant technological industry has grown up to provide prompt and effective servicing.

In parallel, higher reliability is being designed into equipment, which tends to offset the increasing cost of visits by a service engineer, which will be of the order of £30 in urban areas, before any work is started. A volume of reprints of papers on service management published in *Harvard Business Review* (1984) is available, and there are books by McCafferty (1980), Normann (1984) and Wellemin (1984).

1.11.8 Stockholders

Realistic costing of stockholding often shows that the primary cost of material is only a fraction of the overall cost of making it available on demand.

To satisfy moderate and variable requirements it is administratively simpler and cheaper to use the services of a specialist distributor. These businesses now have sophisticated telephone answering systems, and rapid delivery services, so that items can be secured in a matter of hours rather than days or weeks. They are to be found in the fields of small tools, fixings and hardware, raw materials and electronic components. The BSI operates a quality certification scheme for distributors, based on the same principles as the manufacturing Quality Systems Standard, BS 5750.

1.11.9 Equipment rental

The effect of withdrawal of capital allowances in company taxation has been to encourage further the renting or leasing of equipment, which is attractive to a company in these circumstances:

- an operating cost is incurred instead of a capital cost
- this is helpful when the company is expanding
- current business must be 'high added value' and profitable
- there is adequate cash-flow available

The organisations providing these services stress that, although more expensive than self-financing, there are economies in internal administration.

Machinery, office equipment and vehicles are obtainable on leases that vary from one to several years. It is economical to rent testgear for short periods, a useful facility when there is a special task to be carried out.

1.11.10 Quality-systems management

Stimulated by the superior levels of quality achieved by the Japanese, the ACARD Report (1982) pointed the need for quality management schemes, which have since been actively sponsored by the Department of Trade & Industry. The British approach, which will be harmonised with International Standards, requires companies to adopt a quality management system meeting the requirements specified in BS 5750, which covers every function and activity of the firm. The company is required to define its internal procedures in a Quality Manual, and to be self-auditing, and is subject to random checks by the certificating body.

The Japanese system, described by Takei (1986) is called Total Quality Control (TQC), and operates in a similar way. Lomas (1983) and Gillam (1985) have described how British Telecom manage quality, and there is useful information in papers by Girling (1980), Mills (1983), Spickernell (1983), and in a Colloquium on Electronic Component Quality (1986). There is an Institute of Quality Assurance, founded in 1919, when the focus was confined to inspection techniques.

1.11.11 General services

A small or medium-sized business can operate with a minimum staff of key specialists, and contract out for services such as catering, cleaning, secretarial and reprographic work. An accounting firm can service all wage and salary payments, and all general accounting. This way of working can help a growing firm, for example until it can afford to employ a full-time accountant or finance manager: it avoids assuming long-term commitments until the path ahead is clarified.

1.12 Management styles and cultures

1.12.1 The need for awareness

It is a step forward in awareness to be conscious that there are different styles and cultures, and a further step to fit them into some sort of understandable pattern. Then it becomes feasible to operate effectively in another culture, by minimising the mismatch in communication methods and social practices.

We expect to find differences in culture between the UK, Latin America and the Orient, but considerable variations will also exist within UK itself, and within different companies in the same industry.

In the short-term in order to liaise or cooperate effectively, it is expedient to tune in to the style or culture:

> when in Rome, live as the Romans do:
> when elsewhere, live as they live elsewhere
>
> St Ambrose (340–397 AD), in 'Advice to St Augustine'

In the longer-term, the style of a company can be changed, essentially under the leadership of the Chief Executive (see Section 2.7), but the quickest way to implement a change in style is to set up a new team for a dedicated purpose, either as a project team (as outlined in Section 1.10) or in a new venture (Section 6).

At the international level, technology is now global, and many engineers will at some time in their careers have experience of other countries. To work effectively, as noted in Chapter 5 (5.6) some personal effort must be made on the language front, but it is also necessary and a courtesy, to acquire some feeling for the host country's culture.

In Europe it is relatively easy, as the Scandinavians, Swiss and Dutch are happy to use English, and bilingual collaboration can work, as was proved on the successful Franco–British projects of the Concorde, followed by the European Airbus series. In any project where language is going to be a factor, this must be identified at an early stage, and time and resources made available for familiarisation.

1.12.2 Relating to the culture

Within a particular organisation, there are often one or more individuals who fulfil the role of 'gatekeeper' – they act informally, introduce people, and pass on ideas to those who will use them. This can be encouraged by nominating suitable older hands to serve an 'uncle', in a neutral role to newcomers in other parts of the organisation than their own.

When changes have to be made that affect everyone, a technique used successfully by some large companies is to arrange informal discussion groups, each made up as a 'diagonal slice' of the organisation structure: this is a mixture of senior and junior people, from across the disciplines and departments.

With other cultures, a mutual appreciation of the differences is helpful. Research done at UMIST and reported in the *Financial Times* (1986) assesses the mutual perceptions of Japanese and British managers, working together in UK.

> The Japanese see the British managers as individualistic, with the priority of developing their careers and valuing personal free time
>
> The British see the Japanese managers as group-oriented in decision making, concerned about taking any 'incorrect' attitude

A similar study of an English subsidiary in Latin American would yield an entirely different pair of perceptions, as would that for a group working in the Middle East. For success in any multicultural enterprise a good deal of mutual understanding, tolerance and respect is essential, and the tone must be set by the people at the top.

There is an ailment known as 'culture shock' which sometimes affects expatriates: it is the consequence of a negative attitude that rejects the differences of culture, climate, conditions and catering, rather than the positive approach of attempting to understand them, enjoy, and learn from them.

1.12.3 Changing the culture

The culture of a company can be changed, as Rogers (1980) described: the Plessey Company took a strategic decision of reorientation to meet the dual pressures of rapid technological change and highly competitive markets, and the change in philosophy from a manufacturing to a marketing orientation had a dramatic effect on the managers of the individual business units. Parry Rogers describes how the structure was changed from an earlier bureaucratic pyramid to a 'parent and subsidiary' relationship, and it is interesting that his previous career had been with IBM (UK) Ltd.

There has been a series of biographies recently of business leaders, with accounts of how they have changed companies: one suspects that these are often reflecting personal viewpoints, because the prescriptions are so different from one to another, and each company has its own unique history and characteristics. Also what was best and worked well in one period may not be right for the future: this is very apparent from Geneen's (1985) account of his time and style at ITT, where dramatic changes have been made since his retirement.

There is a lesson to be learnt from the good projects, where a multidisciplinary team work together harmoniously and achieve the intended results. As was done in the studies of 'excellence', if we can observe and identify pragmatically the role models that work well, there is then a sound basis for tailoring future enterprises.

Quoting from a Scottish authority, Robert Burns (1759–96):

> O wad some Pow'r the giftie gie us
> to see oursels as others see us!
> It wad frae mony a blunder free us
> And foolish notion
>
> (from 'To a Louse')

1.13 Reading list

ANSOFF, H.I. (1968) 'Corporate strategy' (Harmondsworth, Penguin) 205 pp.

ARMSTRONG, M. (1986): 'Handbook of management techniques' (London, Kogan Page) 573 pp.

BAKER, J.W. (1985): 'When the engineering has to stop', *IEE Proc.*, **132**, Pt A, pp. 193–198

BIM (1985): 'Improving management performance' (London, BIM) 42 pp.

DRUCKER, P.F. (1954): 'The practice of management' (London, Heinemann/Pan) 480 pp. (a first choice of several books by Drucker available in paperback editions).

HANDY, C.B. (1980): 'Understanding organisations' (Harmondsworth, Penguin) 480 pp.

KNIGHT, K. (Ed.) (1982): 'Matrix management: a cross-functional approach to organisation' (Aldershot, Gower 2nd Edn.).

ODIORNE, G.S. (1979): 'MBO II – a system of management leadership' (Belmont, California, Fearon Pitman) 360 pp.

PETERS, T.J. and WATERMAN, R.H. (1982): 'In search of excellence: lessons from America's best-run companies' (New York Harper & Row), 360 pp. (and in paperback).

PUGH, D.S., HICKSON, D.J. and HININGS, C.R. (Eds.) (1984): 'Organisation theory – selected readings' (Harmondsworth, Penguin 2nd Edn.) 447 pp.

'Engineering Manufacturing Systems' (1988) (London, Peter Peregrinus) – 17 articles by IEE contributors.

Chapter 2

Managing people

2.1 Summary

Managing people is not the exact science in which engineers or scientists have been trained – it is more nearly an art, although practical psychology has a contribution to make. It so happened that some of the definitive work of the behavioural scientists was done in technological industries (Western Electric's Hawthorne works, studies in engineers' attitudes by Herzberg, some pioneer work by Anne Shaw at Met-Vick with their women employees, job enrichment at Bell Telephones and with ICI, and Burns and Stalker's studies of electronic companies in the 1950s).

There is a company's point of view of what it expects from an employee, which may be in harmony or conflict with that person's perceptions and career aspirations: their mutual interests may coincide over a period of time, but will diverge when the individual has to consider job mobility, either spontaneous or enforced.

It is good practice for managers to review and counsel their subordinates, and most companies have procedures for this that are more or less effective. The manager has a further contribution to make in a coaching role, which he should see as a professional obligation: but, in a dynamic situation where everyone is very busy, the long-term benefits of investing this time and effort may be neglected in favour of achieving short-term goals.

An appreciation is made of the contribution in a business of a professional Personnel Department, and also what the individual can do by way of self-development and career planning. A formal training in the sciences can provide a launching pad from which to diverge and make a significant impact elsewhere: in business, the arts, literature, public affairs and statesmanship. In a changing world the specialist can guard against obsolescence and frustration later in life by seeking a wider spectrum of interests, if not in depth at least at the level of 'awareness'.

Whether remaining within one's original discipline, or branching out elsewhere, qualities of leadership will make a contribution, whatever the field of activity. There are 'born leaders' but, like any skills, this competence can be taught and developed. It is particularly necessary in order to be effective in project management, when launching new ventures, and in communicating ideas, whether for project proposals, marketing, public relations or in formal industrial relations.

2.2 Lessons from behavioural studies

2.2.1 Effectiveness by motivation

The training of most engineers has included little reference to applied psychology. A great deal has been done in this field during the last 50 or more years, some of its conclusions contradictory. Most that has emerged now seems self-evident and common sense. General public attitudes have changed, and what is acceptable now was revolutionary in its day.

The earliest contributions towards improving working conditions in industry and business were made by enlightened individuals, motivated by humanistic considerations, rather than in the search for effectiveness and efficiency. This association came later.

An early benefactor was Sir Titus Salt who built a model mill and town complex named Saltaire, near Bradford in 1853. Seebolm Rowntree's concern for welfare at his York factory in the early 1900s, and Mary Parker Follett's work in the Boston area from 1890 to the 1920s laid some of the foundations. She linked an understanding of the human factors underlying organisation with the considerations of efficiency that had motivated F.W. Taylor and Henri Fayol.

It was in the 1920s that systematic work in industrial psychology began, both in linking the efficiency and well-being of workpeople, and in developing techniques for testing, appraising and selecting personnel.

Insights were gained through the classic experiments from about 1925, at the Hawthorne Plant of the Western Electric Company, by a team of scientists from Harvard University, under the direction of Professor Elton Mayo. This led to a better understanding of human motivation, relationships and behavioural patterns: this has been extended relatively recently from the level of the workplace to the behaviour of groups and of middle and higher management.

The original publications are well worth reading: in brief summary the main contributions have been as follows.

2.2.2 The 'Hawthorne effect'

This was identified only in the later stages of a major investigation that continued for several years. Studies made at the turn of the century by F.W. Taylor had shown how to analyse work, so that operations might be performed in the most mechanically efficient way. The original intention of Mayo's programme was to assess the influence on productivity of environmental factors. The main trials were carried out on a group of girls assembling telephone relays, each of about 40 parts, a process already well-developed by method (time and motion) study.

The experiments were planned well, changing one parameter at a time, and using a control group where no changes were made. The team altered, in turn, the lighting levels, the layout, interior decoration, temperature and work breaks.

The surprising discovery was that correlation with productivity was nonlinear: as each improvement was tried performance notched up a step, and then remained high when everything was restored to the initial state.

It was then realised that the favourable reaction of the workpeople was not due to the improvement of conditions, but to the interest that was being shown in them by the management and the team of young Harvard researchers, and the unity that the experience gave them as a work group. For fuller details, see Roethlisberger (1939 and 1942), and summaries by Singer (1972), Armstrong (1984) and other personnel textbooks.

This work has stimulated many more developments in group motivation, notably 'quality circles' by the Japanese in the 1960s, and the Swedish Volvo system of car assembly by a work group, rather than on a conveyor line, in the 1970s.

2.2.3 The dynamics of organisations

While the Hawthorne investigations were going on in the manufacturing wing of the Bell Telephone System, Chester Barnard published 'The functions of the Executive' (1938, reprinted 1968). His career had extended over 40 years with the American Telephone & Telegraph Company, where he started in the statistical department, culminating in the presidency of the New Jersey Bell Telephone Company.

In his book he stressed that co-operation within an organisation depends upon three factors: physical, environmental and social. He identified that the formal organisation structure is supplemented by an informal organisation, which has the potential to contribute greatly to the vitality of the organisation as a whole.

Disregard or disturbance of the equilibrium of these systems can inhibit successful co-operation. Conversely, the strategic factor in co-operation is leadership and moral creativeness.

In the development of leaders this creativity must be awarded a weight comparable with that given to their technological capability. He was among the first to stress the importance of encouraging upwards communication, and contributions from below.

It is said that Barnard's ideas were regarded more as a personal hobby than as corporation policy within the Bell System. But there was a receptive atmosphere reaching back to the early days of the century, when Theodore Vail appreciated that to avoid pressure for nationalisation, and to satisfying the regulatory Agencies, the telephone system must be seen by the public in a favourable light. He coined the phrase 'our business is service': all personnel were trained to provide customer satisfaction, which favoured a spirit of internal cooperation. Drucker (1973) has developed this theme in his chapter 13.

2.2.4 Quantifying 'management style'

Earlier explanations of behaviour in organisations tended to be dissertations on moral philosophy, and so do not become widely appreciated. The breakthrough in general understanding was when Douglas McGregor (1960) put forward a simplified model, to which others added numerical measures.

What McGregor postulated was a scale of assumptions about motivation and human behaviour, between two extremes, which he called:

> *Theory X*: the cynical and pessimistic view that the average person avoids responsibility, dislikes work, and can only be managed by coercion, close supervision and threats
> *Theory Y*: based on the optimistic assumption that people will act responsibly, and that given the chance, most people can make creative and imaginative contributions

Acceptance of McGregor's thesis was impeded by the mistaken assumption that he recognized only the two extreme conditions, *X* or *Y*. He subsequently made clear that he intended to imply that practical situations lay somewhere in between on this scale, and could be influenced to move either way.

McGregor highlighted as his main conclusion that the desirable *Y* objective could be achieved by:

- recognising the needs of the individual
- recognising the needs of the organisation
- creating conditions to reconcile these two needs

2.2.5 Management by self-control

McGregor stated the principle that:

> Man will exercise self-direction and self-control in the service of objectives to which he is committed

Drucker (1954) had arrived at the same thought in his book 'The practice of management', as set out in his chapter 11, 'Management by objectives and self-control'. From this grew the wide acceptance of the principle that persons will periodically discuss with their boss the current objectives, and agree targets for results: then accept the responsibility for achieving them without further over-seeing or supervision.

A number of writers have expanded the theme of 'management by objectives', for example Humble (1970 and 1973) defines MbO as:

> A dynamic system which seeks to integrate the company's need to clarify and achieve its profit and growth goals, with the manager's need to contribute and develop himself. It is a demanding and rewarding style of managing a business

Odiorne (1979) traced 15 years of the development of MbO as a management system, from its initial applications to management performance appraisal. He refers to the policies of decentralisation in large corporations and comments 'MbO made decentralisation work'. Reddin (1971) dealt with the ways of using MbO to achieve organisational change and 'unfreeze' situations. It is effectively a contract between a superior–subordinate pair, repeated at each level in the organisational hierarchy, or between the leader and his team. The emphasis in defining jobs has to be on output, achievement and effectiveness, with quantified objectives, rather than on input and behaviour patterns.

Seyna (1986) of Kodak offers this definition of MbO:

> 'MbO is the system and philosophy that integrates all management techniques and all human contributions into a unified organisation to achieve a common purpose and individual fulfilment'

he derives this from his historical review of how the concept of MbO has evolved over 30 years.

2.2.6 Hierarchy of human needs

The research of Mayo's team in the 1930s had identified that motivation was not entirely a matter of money, but involved other social needs such as job satisfaction and recognition. Maslow (1954) put forward a further scalar model that assists in visualising a fairly complex set of relationships. He identified five categories of human need:

(*a*) physiological – the needs in order to survive
(*b*) safety and self-protection
(*c*) social acceptance and relationships
(*d*) esteem: self-esteem, and esteem in the view of others
(*e*) fulfilment: of one's aspirations and potential

This is a hierarchy or ladder, because the first two primary needs (*a*) and (*b*) must be satisfied before the others (*c*), (*d*) and (*e*) are realised. Once the basics can be taken for granted, the higher order needs become the main motivators.

For half a century analysis seemed to concentrate on the workbench and clerk's desk. Not until 1957, when Herzberg published his behavioural research on professional groups, was there any attempt to look at the motivation of 'knowledge workers'. He studied the attitudes of large numbers of design and technical staff and accountants in the Pittsburg area, mostly in the engineering and chemical industries.

Herzberg introduced the idea that there were separate positive and negative motivating factors. A job could be made more satisfying by enriching its content and responsibility, but motivation will be reduced if what he called the 'hygiene' factors are neglected. He was using the word in the sense of preventive medicine. Knowledge workers and professionals in general both need and expect opportunities for achievement and recognition in their work.

2.2.7 Job enrichment

The idea of enhancing the content of jobs was taken up by several companies in the 1960s, and Volvo's car assembly teams are a well known example. The work groups are self-regulating, and voluntarily interchange their tasks. Quality is enhanced, although productivity remains at about the level of a conventional assembly line. A major advantage in Sweden was that staff loses and turnover were much reduced, where previously it had been difficult to recruit enough people.

ICI have had a substantial enrichment programme in the UK: productivity has been improved with Herzberg's initial assistance, mainly by increasing the span of responsibilities. Extensive programmes were carried out in AT & T with all grades of telephone staff, and in Texas Instruments, and have been reported by Ford (1968 and 1969), and by Paul and Robertson (1970) of ICI. A volume of selected readings on the background and practice of job design has been published in paperback by David and Taylor (1972).

2.2.8 The charting of management style

The idea of McGregor's linear *XY* scale has been taken further into a matrix model by Blake (1964) in his 'managerial grid'. There are scales on two axes, each numbered 1–9, the one representing concern for people, and the other concern for results. McGregor's extreme case, 'theory *X*' is then 1.9 on Blake's matrix, and 'theory *Y*' is 9.1; 5.5 would be a middle-of-the-road management style. The model is used in management training as a vehicle for analysis of current style, and a basis for discussion of how it can be moved towards an ideal of 9.9, combining both concern and the achievement of results.

Reddin (1970) went further, by adding a third dimension to form a '3D MbO' matrix. The additional axis is a scale of 'effectiveness', and is a measure of whether the manager's concern (which is an input) does in fact produce an effective output of results. The emphasis on effectiveness highlights the individual who looks for optimum solutions, as distinct from the busy person who is distinguished by high activity rather than results.

2.2.9 Achievement and excellence

The needs of people at managerial levels were identified by McClelland (1961) as being:

- personal achievement in competitive situations
- affiliation with work colleagues
- power and influence

and he observed that the balance of the three needs varies between individuals according to their personalities, and at different stages in their careers.

A classic study by Levinson (1968, 2nd edn. 1980) dealt with the psychological and social influences on senior executives, and their roles in contempory society as leaders and teachers, an aspect of leadership discussed here in Section 2.7.

A different and more popular style of presentation is used in Jay's books 'Management and Machiavelli' (1969) and 'Corporate Man' (1971), which contain both sound, cynical and entertaining judgments. Even though thought shallow by academics, this kind of writing does achieve very wide readership. Townsend's 'Up the organisation' theme (1970 and 1984) goes further in this genre and can be misleading, as he wrote of his experience in directing a large number of identical car-hire depots, with well defined routines. But these popular books are very readable, as are the host of biographies of business leaders.

A series of books with great impact on the managerial and business community was sparked off by 'In search of Excellence: lessons from America's best run companies', by Peters and Waterman (1982): they set aside academic theory, and identified and analysed pragmatically the key factors. Publication was timely as realisation had dawned that industrial leadership was moving to other nations: in particular, Japan was achieving quality standards in many products, substantially higher than we were accustomed to.

Where academic books achieve sales in thousands of copies, the 'Excellence' book has topped seven million. It has had more impact on more people than all the previous literature. Drucker had the previous records for best sellers: he made no pretensions to quantified research, but is highly readable in an anecdotal way. Case histories are interesting to read, but before applying the lessons to one's own situation, it must be appreciated that some of the circumstances, factors and constraints may be different. Goldsmith (1984) has matched the American study with one on British companies, and several other books have continued the 'excellence' theme.

2.3 Fitting the man to the job, and the job to the man

2.3.1 The job holder's view

Changing opportunities in innovative and high-technology industries, and changing needs of these industries necessitate job flexibility. Individuals should accept as the norm a series of job changes in the course of their career. These should be seen as opportunities, not problems.

As Francis Bacon (1561–1622) said: 'a wise man will make more opportunities than he finds'. Some changes may be forced on one, and others be the result of personal choice – either way, the individual owes it to himself to be prepared to adapt his abilities to changing circumstances.

In a professional career, the initial step is to qualify by studying a specific area of knowledge in depth. Entering new technologies individuals have the opportunity to go on, specialise and innovate in an area currently opening up. Whether structured as a project or not, this dynamic phase has a finite duration. In preparation for the next and succeeding tasks, it is only wise to broaden one's spectrum of knowledge. The first essential is an awareness of subject areas outside one's specialism – entering them in depth is no problem; it is just a matter of allocating time and effort where it will have the greatest effect.

The point is well made by reference to the 'T-shaped' man (Fig. 2.1). He starts as a specialist, getting to know more and more in depth, about less and less of the wide spectrum of knowledge. But, he should keep himself aware of what is going on elsewhere, even if this knowledge is at a shallow level. The worst degree of ignorance is to have 'never heard of it'.

As his career develops (and we must think in terms of the four decades from the 20s to the 60s), the model can be extended with a time axis, as Fig. 2.2. Plotting the path of people's careers, we will find that at some point someone may have achieved greatness, but for others it was thrust upon them, as an unforseen responsibility, or promotion.

The T-shape depicted refers to the *knowledge* spectrum: another could be constructed for the individual's *abilities* portfolio, which might include being observant, good with people, leadership qualities, literacy, ability in compiling reports and at public presentation, self-organisation, and energy and vitality.

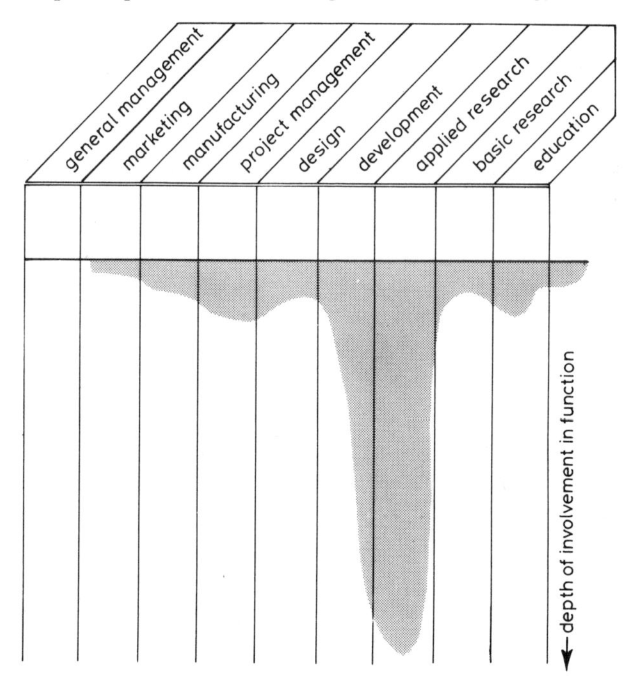

Fig. 2.1 ***The 'specialist's' areas of awareness profile***
The 'T-shaped' man

A person may be born with some of these attributes, but can cultivate the others through hard work. We must not ignore the lessons of show business or professional sport: a couple of hours of high activity will be preceded by preparation and relaxation.

The initiative for adapting to the work environment should lie with the individual. Support can be expected from the education system, the professional institutions and the employing organisations. Probably less is done formally in UK than elsewhere, towards the 'formation' of individuals, and their 'management development'.

There are pluses and minuses in this: the formation process is strongly defined in some countries, and this may result in career patterns that offer less freedom and flexibility for changing personal preferences, and response to events.

One aspect of this freedom is the provision of opportunities for 'career breaks', which should be a recognised option. These can be exercised in a number of ways:

- secondment to another sphere for a period, for example from industry to teaching, to the Civil Service or other public office
- 'intrapreneurial' opportunity, to develop a new product or service as an independent unit, within the main parent organisation
- exchange with an associated organisation overseas
- postgraduate course, technical or Business School
- voluntary service overseas in Third World countries
- hospitalisation, or care of young children

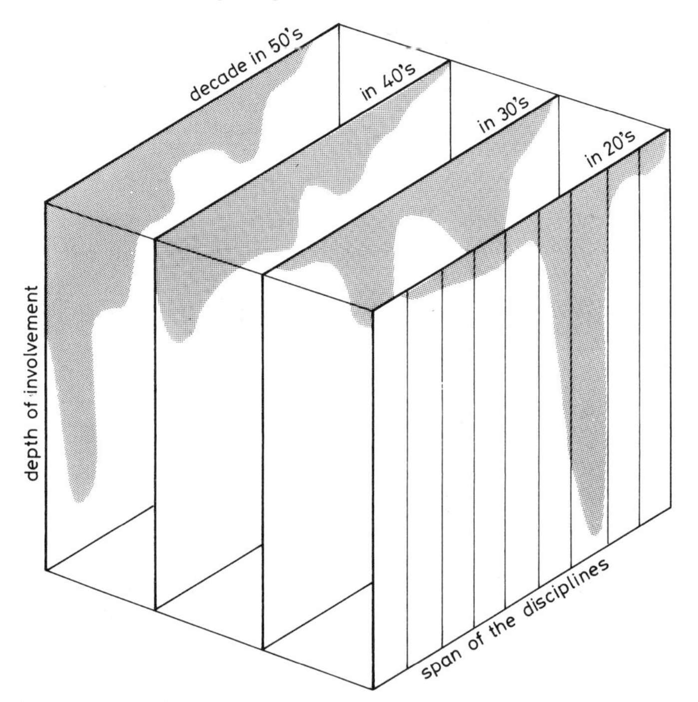

Fig. 2.2 *As a career develops towards 'general management' in the third and fourth decades, a person gets to know less and less about more and more, as the profile of work responsibilities changes and broadens*

The IEE have issued a Professional Brief (1985) on 'A study of members' career breaks'. It includes six personal profiles, and deals with the problems of re-entry. It was based on a sample of about half of the Institution's 1300 women members, but the concept of 'career breaks' has wider significance for the total membership of over 80 000.

The Engineering Management Society of the IEEE (USA) has issued a helpful book by Gray (1979), titled 'The engineer in transition to management – a learning tool for the engineer or the professional newly promoted to management'.

There are many new opportunities for professional people of mature experience, as they approach mid-career, but they must be prepared and willing to apply new concepts, face new challenges, and assume new personal roles.

The IEE Professional Brief (1986) on 'Career Development' defines (in an Appendix) six levels of responsibility, and indicates statistically the proportion of the membership achieving them: corresponding remuneration is identified in the IEE's annual Salary Surveys. Similar data for the USA is published by the IEEE.

2.3.2 The organisation's view

Only a proportion of the larger employers are active in preparing people for their present and future jobs. These firms carry the burden of training staff who then go to other employers. Within reasonable limits this can be advantageous, as the primary employer has the opportunity of 'first pick' of those he trains, and can select and encourage those who will be most useful. Some wastage will be allowed for in the process of long-term manpower planning that will be found in many big organisations, as outlined in Armstrong (1984, chapter 10).

For a period of time, there will be parallel paths and common interest, between the employee and the firm. This is unlikely to apply always, and it is natural that their paths may diverge. This happens when a team comes together for a project, and then disperses when it is finished.

Outstanding individuals will manage their own careers, and rise through a large organisation, or be mobile and pass through several, or strike out on their own as entrepreneurs, as in the case histories in Section 6.

There are a number of things that an organisation can do to ensure that it makes the most of the talent it has:

- avoid undue structural rigidity, so that jobs can be adapted to the talent available
- rearrange organisation structures, or the make-up of divisions or subsidiary companies, to match the strengths and weaknesses of the available talent, only augmented from outside with key people, if absolutely necessary
- restructure very large units into a more manageable size of not more than 200 or so, so the unit chief executive has short lines of communication with all personnel, and can know and be known by virtually everyone
- if a job gets too big and a replacement is needed, split it into two: alternatively, combine two jobs into one to stretch an able individual
- identify the 'flyers' or 'fast-lane' individuals, and broaden them with a succession of posts in different functions, and at different locations
- make available opportunities sufficiently challenging that few are tempted by 'head hunters'
- identify personal training needs, structure programmes of training opportunities, and use both internal and external resources for management development
- be prepared to budget a sum for personal development comparable with that devoted to technical development

Internal training is not only achieved through formal teaching. Within the larger organisations much benefit can come from setting up 'task-forces' to examine particular areas of problems and opportunities, arrive at recommendations, and present these to top management. Individuals learn from operating in a multi-disciplinary team, and there is cross-fertilisation of ideas, coupled with the sharing of experience of mature people in what amounts to a 'teach-in'. The motivational results are excellent, and any recommendations are already pre-sold to those involved, in the Japanese manner of consensus management.

2.3.3 Separating the man from the job

There are fewer jobs at the top, so the majority can look forward to promotion only when someone moves on, or through expansion. 'Peter's principle' recognises that people get promoted beyond their competence – an excellent research or sales person may be a disaster as a manager. A corporate policy of moving people sideways is preferable to leaving them too long in one position.

Unfortunately, from time to time, organisations are overtaken by events, and have to close or dispose of subsidiaries, and declare widespread redundancies. In the USA this tends to happen more often, but can be less traumatic, as the closure of a branch plant tends to be regarded as similar to the conclusion of a project: domestic mobility is still an acceptable option. This is probably a carry over from the pioneer days of opening up the West, for full industrialisation on the Pacific coast was as recent as the Second World War period.

When an employer has to cut back, there are several options open to professional employees:

- business decline rarely occurs overnight – some people will be aware of the negative factors, and diminishing prospects, and will make a planned move before it is forced on them
- the statutory redundancy payments will be available to those who stay till the end and, typically, provide a 'breathing space' of 6–12 months
- 'outplacing agencies' can provide advice on personal redeployment, at a fee cost: in some redundancy programmes this is covered by the employer
- a wide range of services are provided by the Professional Employment Register (PER) of the Department of Employment, and the Manpower Services Commission
- where a company is closing down or cutting back there is often the opportunity for a group of the staff to make a buyout offer for a part of the business, to which they are committed (see section 3.8.3, Finance)
- self-employment is a viable alternative

The overhead costs to a company of employing an individual are relatively high, and may effectively double his direct salary or more. It is attractive to companies to employ individuals as subcontractors as and when they are needed, where the individual can provide specialised services and skills, or act as a consultant.

The employer will often offer a limited contract to a departing member of staff, and that buys time while the individual locates additional clients – the tax treatment of a self-employed person is quite favourable: he will have to register for Value Added Tax (VAT) in the first quarter that his annual turnover rate exceeds the current minimum level, adjusted in each Budget, and currently at about the £20 000 gross level.

2.4 Review, appraisal and coaching

2.4.1 Type and frequency of review

Regular review of corporate objectives, budgets and results is an essential feature of operating a business, and is outlined in Sections 1.6 to 1.8. Fig. 1.6 illustrates that there will be a two-way flow of information: contributions upwards to the formulation of the strategies and operating plans, and expansion in detail as instructions proceed downwards, in order that the unit managers or supervisors know what is expected of them.

The structure of working relationships at each level in the hierarchy involves the level above, and the level below:

(the manager) + (his boss = 'father') + (boss's boss = 'grandfather')

In making an analogy with the family, hopefully family tensions are not implied! None the less, between a subordinate and his superior there is a complex working relationship, comprising both the impersonal company aspects, and the personal attributes and contributions of both parties.

One feature of the management scene in the USA, often noticed by Europeans, is its professional style. A high proportion of people have experienced some form of Business School training, and they share a common approach and shorthand vocabulary.

Periodic reviews of the company's work inevitably raise aspects of personal performance, and it is best for both parties to identify these elements separately. The matrix is a simple model of what is involved:

mode / subject	casual contacts	progress review	planning
	short term	medium term	long term
The job	(*a*)	(*b*)	(*c*)
The person	(*d*)	(*e*)	(*f*)

Explaining in more detail these six relationships:

(*a*) The short-term, day-to-day contacts help the manager to keep the job going on the right lines, following the principle of 'management by exception' – if something unforseen occurs he will seek the boss's guidance.

(*b*) Regular formal reviews of the progress of the programme of work are necessary at medium-term intervals, say weekly or monthly, and these are preferably minuted with agreed action and names, either as one-to-one reviews, or by a boss with his team.

(*c*) Long-term programmes typically will be put together for the year ahead, and possibly projected up to 3–5 years and rolled forward annually: these provide the framework for the more frequent medium-term reviews.

(*d*) Subordinate–superior pairs will be likely to have informal contact, and more frequently at supervisory levels – in the process they get to know each other better and thereby improve the effectiveness of their interpersonal relationships: they understand each other in terms of getting the work done. An effective boss casually makes a point of having a word with everyone in his unit from time to time (possibly with a pocket reminder system), and will be seen about and interested in what they are doing: this technique can work very well but begins to break down as numbers in a unit approach a hundred or so.

(*e*) At reasonable intervals every (subordinate–superior) pair should discuss the personal aspects of the job: a quarterly or half-yearly review is less formidable than a very formal 'annual review'. The medium-term review will look at previously agreed targets, whether they have been met, and what assistance and guidance may be needed – it is not effective for a boss to make a personal review with a subordinate on a casual basis: it should be a formal occasion at pre-arranged intervals, and an agreed record kept.

(*f*) An annual review with the boss should be in greater depth: this is the opportunity to take the longer view of the subordinate's aspirations, to give an opportunity to express preferences for alternative future tasks or projects – the boss will try to help identify and agree what further training may be desirable. At more junior levels this may relate to technical subjects, and for people moving into supervision and management there should be adequate preparation, so that they can more effectively execute present and forseeable tasks

The above outline is in terms of middle and senior management in the organisation hierarchy. The same pattern in simplified form should extend down through every level, where the targets will be shorter-term. There is the useful principle of 'time span of discretion', originally put forward by Jaques (1965, chapter 8). For example, the Chairman of a company is likely to set the pattern for the next 5–10 years; the Managing Director is more concerned with the current year's operations, and planning for next year. A Divisional Manager may be reviewed monthly or quarterly, and a Departmental Manager rather more frequently. A Supervisor is largely responsible for day-to-day targets in the current week, and a skilled operative is responsible for the job in hand.

It is helpful to recognise that there is also a 'time scale of abstraction'. At the operational and supervisory level there is (or should be) very specific knowledge of what is to be done. At the other extreme, the Chairman has to dream dreams for the future, then guide his team towards exploring their potential.

On promotion, a manager must appreciate that he will have to think on a different and less detailed level than before, at least for the strategic aspects of his new job. There is a pitfall here for the smaller and growing new ventures. The 'Directors' are for 95% of their time busy executives or technical specialists. When they are really acting as Directors and making strategic decisions as a Board, they must remember that they are 'wearing a different hat'.

2.4.2 Constructive criticism

In an ideal relationship between superior and subordinate, a positive approach is made to the formal review. The superior's role should be to assist, to build up the subordinate's strengths, recognise his need to feel that he is performing well, and help him to meet this need. Negative aspects will come to the surface, but the boss's approach should be to build on the strengths, with a minimum of negative criticism and disciplining of weaknesses.

A company should ensure that these one-to-one reviews are made between every pair in the hierarchy. People unused to conducting an objective appraisal are nonplussed by the task and, in some organisations, it falls into disuse for this reason. Knowing how to conduct an appraisal interview should be part of the training that every manager receives. The fact that it is company practice to make personal reviews should minimise the embarassment between people who already have a close day-to-day relationship.

Details agreed between a manager and his boss are primarily their mutual concern, but should be put on record for reference at the next review. This is where company practices differ – one method is for the manager to write a memo (against a check-list of standard headings) recording his understanding of what was agreed.

Large companies tend to have a defined procedure with printed forms, which may run into a number of pages. Armstrong (1984, pp. 174–6) gives simplified examples. Although the process then becomes more mechanistic, it does ensure uniformity of application. The Personnel Department usually monitors the frequency and content. This can provide the feedback of training needs, that can be used to build up a company training programme.

Appraisals are linked, indirectly or directly, to salary reviews and considerations of promotion. Here the boss's boss ('grandfather') has an important part to play, both in taking a strategic view of manpower recources and in talent spotting, but also as a moderator of the judgments and prejudices of the direct boss of the person assessed.

The trend towards 'result oriented' management lays the emphasis on the achievements by persons, rather than their qualifications and characteristics. They get results in their own inimitable ways. For purpose of recruitment and selection, a characteristic profile may be useful as a means of sorting and ranking applicants, but only as a preliminary for face to face assessment.

Job descriptions are a key element in the design of organisation structures, and provide an initial brief for a newcomer to a job. If too prescriptive and detailed they will inhibit initiative and change, and should be regarded as a useful tool, to be adapted whenever circumstance change. They should not define duties in detail:

the appropriate place for such instructions, if they are essential, is in procedures manuals, where they are more readily co-ordinated and updated.

2.4.3 Coaching

Adair (1983) and others claim that successful leaders are teachers, although the converse is not true, according to George Bernard Shaw, who said that 'he who can does, he who cannot, teaches'. It is generally held that to prepare for his own promotion a manager must ensure that he trains his successors.

The companies that have been assessed as 'excellent' are those that have been strong in training, personal coaching, and in passing on to the younger members the concepts and principles on which the founders built their success. Particular examples in high technology, picked by Peters and Waterman (1982) were:

Digital Equipment	Data General
Hewlett Packard	Intel
IBM	National Semiconductor
Sclumberger	Raychem
Texas Instruments	Wang Laboratories

Goldsmith and Clutterbuck (1984) made a similar study of British companies and chose, among others, these technological companies:

BOC Group	Plessey
BTR	Racal
GEC	STC
NCB	

The common factor has been that these companies have had continuity of leadership over long periods, in some cases from the founders and their families, or by the person who has carried through a major restructuring. The leaders have been visible and in personal contact, at least at the more senior levels. In a large Group this role is assumed by the senior man at each site.

The orientation towards people is termed at Hewlett-Packard 'management by walking about'. In most of the companies there is an informal/formal or 'loose/tight' style of management. The units are small enough for personal contact and simplicity of administration, yet results are monitored effectively from the centre.

A helpful chapter on 'Coaching and counselling' by Edwin Singer appears in 'The Experienced Manager' edited by John Humble (1973); also in Singer and Ramsden (1972). Bosses must manage their own time effectively, so that they can devote a substantial part of it to assisting subordinates, rather than delegating the role entirely to professional trainers. Much of this time must be reserved for perceptive listening. The boss can coach in the context of the job itself, on the job, rather than by abstract generalisations and examples from other unfamiliar fields.

He gets to know each person, and can adjust his approach to their needs, attitudes and state of motivation. Coaching by a superior is an investment of his time, which will pay off as the subordinate grows to take over some of the boss's responsibilities and work load.

Some companies adopt the device of an 'uncle', an experienced member of staff to whom younger people can turn for advice, outside their own department. Sir Peter Walters (1986), who is quoted again in Section 2.10, speaks of such a person in British Petroleum, who was mentor to a whole generation of 'bright young men'. An allied role is that of *gatekeeper*, someone people turn to for specialised information (Taylor, 1975).

2.4.4 Management by objectives

The concept that a manager has self-control of the responsibility delegated to him was stated by Peter Drucker (1955, chapter 11). Once objectives have been agreed with his boss, it is up to the manager to go and achieve them. This quite simple idea has been interpreted and implemented in different ways, and it is estimated that half of the large US companies in the *Fortune* top 500 use the approach in some form or another:

- as a basis for management development at the level of the individual manager, according to John Humble (1970)
- as a complete system of management, according to George Odiorne (1979)
- as the basis for an annual performance review, reported to the Personnel Department
- as a punitive target-setting device, as practiced by 'hard' managements

Where the practice of MbO has fallen into disrepute, it was because it was employed to discipline managers who failed to reach targets, rather than as a vehicle to help them to overcome difficulties. In other applications, failure has been due to a proliferation of paperwork, with perhaps 20 pages of standardised documentation considered at each cycle of review.

The original concept of MbO was *not* to include every item of detail that might be found in job descriptions and work instructions, but to apply the Pareto 20/80 principle – attention to 20% of the factors will produce 80% of the results. It is best to agree objectives for perhaps half a dozen key areas of results, which is another application of the 'management by exception' principle. The less critical areas of results can be delegated, and in turn become the 'key areas' for subordinates.

The other feature highlighted in MbO is that the target results in each key area are expressed quantitatively. This may require some thought and ingenuity, but in the process the problem becomes better understood.

For example, instead of the instruction that a telephone switchboard operator 'should answer all incoming callers promptly', a quantified performance standard is set. I.H. Cohen (1986), Managing Director of Mullards, has stated that, to increase the quality of customer service, the switchboard performance standard of answering 75% of all calls in 15 seconds was raised to 90% in 10 seconds, by involving the staff in an improvement programme. The time for processing orders was improved from 70% in 48 hours to 90%. This was one element in the comprehensive quality improvement programme of the Philips Group, described by VAN HAM (1986).

The elements of MbO are now so widely known that it has become absorbed into normal good management practice, instead of being identified as a special technique.

Professional engineers have an edge when they enter management, for they have three inherent advantages:

- their whole training is 'results oriented'; they are practised in making things happen, and creating things that are effective and work
- they have a numerate rather than an emotional response to problems, their analysis, and their resolution
- they are ingenious, and likely to come up with original and even outrageous solutions

To quote again from Harry Tomlinson (see chapter 1, page 26).

> No one can practise Druckerism so well as a result-oriented engineer, who is not afraid to break the rules and ignore taboos

2.5 The personnel function

2.5.1 A professional approach

In a large company, a technical manager will be aware of his professional colleagues in the Personnel Department, and will depend on them for many aspects of personnel management, and the service they provide, within the framework of the company's policies and procedures. But he should remain the primary point of contact with people in his own team, and not delegate this responsibility to Personnel, whose correct operational role is the provision of support services and advice.

At the policy-making level in a company, the Personnel function ought to be represented on the Board by an executive Director (who may also have other responsibilities). For large manufacturing organisations industrial relations is potentially a critical area, if not handled wisely at top level. In a high-technology company, the know-how and goodwill of the technical staff must be preserved as it represents a considerable capital asset, expensive to replace, whether by further recruitment or by training.

A particular difficulty in a young growing company is that, in the early days, the few key people at the top are very busy. Despite this, they need to acquire an awareness of the basic obligations of personnel management, some of which are statutory requirements. If they were previously with a large company they tend to either follow, or react against, this prior experience.

So long as a team is small, close personal contact and common sense help to minimise difficulties. At about the level of 100 staff, it becomes essential to have someone knowledgable enough to shoulder the administrative side of the personnel function, possibly doubling up with other work. As a company grows, a personnel professional should be recruited to the management team.

A growing company, without a formalised personnel function, can seek advice from the local Industrial Relations Officer of the Department of Employment, particularly for assistance at the supervisory level. A technology-based company will usually engage and consult its professional staff through its senior managers, with back-up from the personnel manager.

The several functions of management have each sought professional status, and the Institute of Personnel Management provides a path to a recognised qualification via examination. It recommends approved courses of study. Originally it was founded as the Welfare Workers' Association in 1913.

Other early developments came from the experiences of industry in the First World War, leading to the formation of the Industrial Welfare Society in 1918 (now the Industrial Society), and the Institute of Industrial Administration in 1919 (merged in 1951 with the British Institute of Management).

The focus of attention has expanded from the original concept of 'welfare', to cover all aspects of resourcing, industrial relations, management and development, at all levels. This replaces the benevolent approach of the better family-owned companies in the 19th and early part of the 20th century.

The history in the public service followed another path, but practices have converged in recent years. Traditionally, the Civil Service was very different to the outside world of industry and commerce, although major institutions took the Civil Service staff conditions as a model.

The first development of interest in management organisation and methods was in the Treasury in the early 1940s. Reforms in the late 19th century had set the pattern for many years, and were based on the Northcote–Trevelyan Report of 1853. This ensured that appointments were made on merit, rather than by influence. Appointments were made into strictly defined classes, and careers followed predictable lifetime patterns of progression through structured grades.

The Report of the Fulton Committee in 1968 recommended setting up the Civil Service Department to take over this responsibility from the Treasury, and to develop further the personnel maangement function, and facilities for training in management.

This change had great significance for the large number of engineers and scientists employed in Government Departments and kindred organisations, where there had been a parallel structure of technical and administrative grades. This was also the time when, under the Post Office Act of 1961, the telecommunications and postal staffs ceased to be members of the Civil Service.

2.5.2 The personnel function in small and large companies

A characteristic difference between a small and a large firm is that the latter will have a professional managing the personnel function. In a small company the obligatory duties will usually be coupled with the administration of payroll, and the terms of employment looked after by the Company Secretary. During growth there is likely to be a critical point when the presence of a personnel professional becomes overdue.

The head of a small unit may be the one who takes an interest in the individuals and their potential, but he may be too busy or not have the inclination. In a well-run large company, there are laid-down personnel practices, and unit managers are expected to follow them. One of the requirements for promotion to a position of responsibility should be that the candidate is given an introduction to these obligations.

The companies that stood high in Peter's (1982) and Goldsmith's (1984) 'Excellence' leagues have been those where these obligations were taken very seriously. There is also a clear trend towards putting the responsibility for personal contact upon the unit manager or superviser, but requiring them to follow the ground rules of a central 'Establishment' function.

There is also an egalitarian trend to 'single standard' conditions of employment, and away from the formal 'classes' of the public service, and the multi-tier grades of large organisations. Now progressive companies offer similar staff conditions and benefits to what were separate 'hourly paid', 'clerical' and 'staff' grades. The most obvious evidence of this is a single and attractive canteen available to all personnel. In the past, some companies operated canteens and dining rooms at as many as six or more levels.

It is easier to adopt these new standards in a growing high-technology company than in a traditional industrial complex, because software and systems engineering, in an office environment, has taken the place of traditional heavy-manufacturing facilities.

2.5.3 The work of a Personnel Department

A 'body of knowledge' in this field has been defined by the Institute of Personnel Management, under these main headings:

- the basis of Personnel Management
- organisational behaviour
- employee resourcing
- employee development
- employee relations

An engineering manager needs to be aware of these activities, without necessarily going deep enough to take the qualifying examinations. A useful recent book of reference is the 'Handbook of Personnel Management Practice' by Armstrong (1984), which, in its second edition, follows the structure outlined above.

It also contains details and examples of generally acceptable procedures, which could usefully be adopted in the early stages of a new enterprise, and which might otherwise be overlooked.

There are a number of relevant 'checklists' in the series issued by the British Institute of Management, which are a guide to good practice:

Checklist no.

6	Manpower planning
7	Induction
8	Filling a vacancy
9	Selecting staff
25	The Employee handbook
43	Recruiting and training graduates
49	Introducing flexible working hours
83	Employee participation

Industrial relations is an area where prior experience is very desirable. Armstrong's book is helpful (chapters 22 and 23), and there are these BIM checklists:

4	Productivity agreements
64	Preparing for bargaining
71	Negotiating a closed shop agreement
84	Preparing for Industrial Tribunals
85	Handling a claim for Trade Union recognition
86	Handling redundancy

A positive approach towards industrial relations is to ensure good communication at all levels, supplying adequate information through supervision, and providing for joint consultation. Any problems will then be identified at the earliest stage, and management should take the initiative in seeking their resolution.

The legal obligations of a business are outlined in Chapter 4.3.

2.5.4 People as a resource

Some managers are good with people – others who have been thrust into management, as a consequence of their technical achievements, may be 'tone deaf' and insensitive in their interpersonal relations.

In technical (as in artistic) fields, talented staff are a precious resource that must be nurtured, if they are to perform well, whether under a technical manager or an artistic impresario. Those already in management, or aspiring to be, must acquire a feeling for 'what makes people tick', either instinctively, or by the example of others, or through some appreciation of what has come to be known as behavioural science.

The terminology of this subject has tended to distance the industrial psychologists from practical managers. Some recent books have helped to bridge the gap by chronicling what happens in actual situations; for example:

- Kidder's (1981) 'The soul of a new machine'
  An inside view of the project team in an intensive programme to develop the first 32 bit 'supermini', at Data General Corporation
- Grove's (1983) 'High output management'
  The President of Intel Corporation provides an insight into how a successful semiconductor company has developed its people

- Auletta's (1984) 'Art of corporate success'
  How Sclumberger Ltd has dominated its market in oilfield exploration by developing its engineers

In the more dynamic technological situations, people tend to be project-oriented; they stay with a team for the duration of the task, then move on, either inside or beyond that organisation. The traditional wisdom is that 'a rolling stone gathers no moss, but it sure gets experience'. Much can be learned from opportunities to experience the inside culture of several different companies.

Relatively few companies have a policy of progressive development of their people-resources, on the basis of a long-term investment. The outstanding example is IBM. Rodgers (1986) has provided insights into IBM's corporate style, which derives directly from principles laid down in 1914 by the founder, Thomas J. Watson senior:

- the individual must be respected
- the customer must be given the best possible service
- excellence and superior performance must be pursued

Watson had worked for the National Cash Register Company, which itself had progressive policies. He set up his own business, initially marketing butchers' scales, time clocks and punched card machines, for which he had acquired the patents. He saw the benefit of raising the status of his salesmen, coupled with quality of the product and the service he provided to his customers.

This tradition was maintained when electronic computers were introduced in the early 1950s, and the same principles were applied to the greatly expanded staff of technical and computer systems people, engineering and R & D staff.

IBM has expanded internationally, to over 400 000 employees and, unlike many companies in the high-technology industries, has a stable record of long-term employment. Internally, however, there is a good deal of mobility in jobs, between locations and between functions, particularly among individuals marked for promotion. Peach (1983) has written about their employee-relations practices.

Much attention is given to training programmes, which typically average 40 hours a year. On appointment to a managerial level, some 80 hours of preparatory training is provided, much of this related to staff management and counselling. In terms of conditions of employment, the company is willing to pay a premium and provide awards for excellence and achievement. This is coupled with the presence of a good deal of peer pressure and mutual criticism, pitched at a professional rather than a personal level.

It is helpful to be aware of some other recent reviews:

HANDY (1984): 'The future of work – a guide to changing society'
GARNETT (1983): 'On the scope for creativity and vocation in industry'
CLUTTERBUCK and HILL (1981): 'The re-making of work'
HICKMAN and SILVA (1984): 'Creating excellence'
PETERS and AUSTIN (1985): 'A passion for excellence'

TOFFLER (1970): 'Future shock – the ways we adapt to change'
NAISBITT (1982): 'Megatrends – 10 new directions'
DAVIS (1987): '2001 Management – managing the future now'
PETERS (1987): 'Thriving on chaos'

(The full references are given in the Bibliography (Chapter 7.2)).

2.6 Self-development and career aspirations

2.6.1 Acceptance of change

It has been said that every problem is an opportunity in disguise. Careers may seem to be disrupted by business reorganisation and redundancy, but this may turn out to be the right moment for the individual to seek a change of direction. Up to this point, your employer's Personnel Department has managed your career – now is the time for you to manage yourself.

It was only some 20 years ago, in the Reith Lectures on the impact of automation, that Sir Leon Bagrit (1964), as he then was, was among the first to say that in the future we must all face the probability of several career changes in the course of a working lifetime. There is a sort of entropy in operation, whereby it is a natural process to move downward from the heights of specialism to become a generalist, and this progression is compatible both with the progress of technology, and the career path of the individual.

If a specialist continues for too long, he gets to know more and more about less and less of the whole spectrum of knowledge, and is stranded in a deep dead end, if his specialism dies.

The personnel departments and management development programmes of the large organisations enable them to manage the individual, and mould the person, in the interest of the firm. It remains up to the individual to manage himself in his own interest – to anticipate and prepare for change, rather than be surprised and overtaken by events.

The prospect of a life-long progression through the levels of an 'Establishment' is no longer a valid career model, even for those in the Public Service. That this became a common expectation in UK, during the period of Victorian and Edwardian stability, is an historic penalty. It has coloured British attitudes, while in the developing world others were opening up new frontiers.

In his own interests the individual should from time to time think through his personal strategic planning, contingency plans and risk analysis. A decision tree can be sketched out, stretching into the foreseeable future: the unforeseeable events may fall within the diverging branches of this tree or outside them. There is a helpful professional brief (IEE 1986) on 'Career Development', which expands on this theme.

In his first play, 'Dangerous Corner', J.B. Priestley (1932) illustrated dramatically how quite small events can launch us upon divergent paths. In his plot a group of people attempt to unravel the events that led up to a tragedy, but in the final Act the author repeated the first Act with a different outcome.

This illustrated that life is random, and that difficult situations and dangerous corners can, with care, be avoided. In an interview some 50 years later Priestley said 'I'm a canny chap, I avoid dangerous corners – I avoid people whose values I don't like'.

Both in fiction and in personal experience, luck (bad or good), and lack of perception or alternatively a brief insight, have marked the turning points in individual careers and in major events. Chance plays a major role in whether we fail or succeed, but, by developing insight, the intuitive processes are enhanced, and the chances of success thereby increased. The process is sometimes called 'gut feel'.

2.6.2 Preparation for a career

Classical studies of Greek and Latin language and history are traditionally thought to be valuable in themselves, not for their contemporary relevance, but to train the mind.

A background in science and engineering is at least equally effective for this purpose. One learns the disciplines of scientific method and numeracy. But it is also desirable to acquire some sensibility to the humanities, and early periods of industrial work experience are an excellent introduction to the field of human relations.

A classical scholar who finds his way into business does not reject his background. Unfortunately engineers and scientists do tend to drop out of their professional affiliations if they move away from practical work. Within the framework of professional qualification, there should be provision and encouragement to continue in their professional association.

For example, engineers who, at the 'student' or 'graduate' level of membership, pursue opportunities in technical selling and liaison frequently lose their right to professional membership.

This is unfortunate, because engineering products increasingly are sold as 'black boxes'. Sophisticated components must be explained to customers by knowledgable field engineers. Products such as the wide range of integrated circuits, microprocessors and memory chips, the great variety of transducers, actuators and hydraulic control components; all must be sold with a high quality of technical back-up.

2.6.3 Divergent careers – a case history

Starting their career in applied technology endows people with the thought processes to make significant contributions in other fields. An interesting case history of a diverging career path is available in the autobiography of Eric Ambler OBE (1985), best known for some 20 novels and 16 screen plays (some of them adaptations for the screen of novels by other people). He is said to have changed the whole character and direction of this genre of literature and film making.

What was different about Ambler was a background unlike that of most literary persons. While at school he built radios, in the pioneer days of Marconi's broadcasts from 2MT at their Writtle Laboratories, near Chelmsford. He operated an unlicensed Army-surplus transmitter, bought off a barrow in London's Farringdon Road Market. He and a friend had the use of a home workshop, and also set up a chemistry laboratory. On the advice of family friends who were professional engineers and members of the IEE, Ambler sat the scholarship examination for the Northampton Engineering College, in St John's Street, Finsbury (now The City University), and came top among 200 applicants.

He started on the B.Sc.(Eng.) course, which then included comprehensive workshop and drawing office training. Being ahead of the lecturers in some of the subjects, he began to spend a good deal of time in the IEE Library at Savoy Place, reading widely. He also began to attend in the public galleries of the Law Courts nearby in the Strand, and also matinee performances at the adjacent theatres.

In the remainder of his spare time he also joined the Territorial Army, and his studies were interrupted when he was called up for service during the General Strike of 1926. On return to College he found it difficult to settle down, so through an introduction obtained a job at the Ponders End works of the Edison Swan Electric Company.

He was the first 'technical trainee' to be appointed, as distinct from the apprentices under contract. He was introduced to the company by going from one shop-floor to another, and from task to task. He found that those tasks that were said to need months to learn, he could pick up in a day. He spent a period on the manufacture of electric lamps and thermionic valves, both quite highly automated processes at that time (1928). His chemistry and mechanical training had prepared him to be an effective trouble-shooter of problems with glass pinch seals and wire drawing.

He had periods in the switchgear machine shop, with the manufacture of storage batteries and their installation on site, and in the dry battery shop; also in the assembly of radio components, where he gained some experience with quality control.

At Head Office, good reports were received, and he was sent for six months to the cable factory in the Forest of Dean. That summer he began a novel based on his father's experiences as an amateur entertainer, an interest shared by several of his relatives. This first essay was never completed.

The company, with two others, was being reorganised as the Associated Electrical Industries Group (AEI). A central publicity department was set up by Miss Nora Miller, a pioneer member of the Electrical Association for Women. Ambler was transferred to work for her at the Head Office. He was given a variety of tasks, such as writing imaginative publicity for slow-moving products, and co-ordinating AEI's stands at major national exhibitions. The Chairman of AEI also required the staff to 'volunteer' to canvass in the General Election for Winston Churchill, as MP of Epping. Ambler was by now aged 21.

It was found that the advertising prepared by AEI's Agency for technical products, to be published in the trade press, often had to be rewritten. Ambler had the background knowledge of the products and the imagination to do this well.

Eventually it was decided that he should transfer to the Agency's staff, to work on the copywriting for AEI.

This marked the first point of divergence in Ambler's career: he also prepared effective copy for a babyfood account, and in the promotion of a chocolate laxative. The colleagues he mixed with were aspiring novelists, playwrights and artists, and he had opportunities to travel in Europe.

In 1935 (at 26) his first book was accepted for publication, a thriller 'The dark frontier'. It was only a modest success at the time, but showed some prevision in its plot about a group of scientists in Ruritania preparing an atomic bomb. A reviewer wrote that 'he has knowledge and speed'. Ambler says that, in anticipating the production and use of the bomb by a decade, he drew heavily on his earlier reading of applied physics in the IEE Library.

By now his 'day job' was as a Director at the Agency. His next dangerous corner was to give up this job after the fourth book, and take the chance of existing as a writer. His publisher came up with a contract for six more books, and also film rights were sold.

When the war came in 1939 he joined up and became an instructor in a Driver Training Unit, and went on to be an anti-aircraft gunnery officer. He was sought out and transferred to the Army Film Production Unit because of his reputation with film scripts, and worked on training films and wartime documentaries. He completed his service in 1946 as a Lieut. Colonel, Assistant Director of Army Kinematography, at the War Office in Whitehall.

In 1947 he wrote and produced a film 'The October Man'. But this was both the beginning and the end of his career as a film producer, and Noel Coward advised him to 'forget films, write more books', which he has continued to do with great success.

2.6.4 Synergy in career experience

Ambler had enjoyed five quite distinct careers; engineer, advertiser, Army, film maker and as an author, and his perceptions of each were enhanced by his experiences in the others, in a synergistic way.

It does no harm for professionals in the applied sciences to broaden their interests. If they remain in the mainstream of their profession, they are better able to become influential interfaces in high places.

Among world statesmen, only a few have a background in applied science. There were two American Presidents, Herbert Hoover, a distinguished civil engineer, who had the misfortune to be appointed when the great depression struck, and Jimmy Carter, who had been a nuclear engineer in the Navy: it was said of him that his fault was that of giving too much attention to details.

Benjamin Franklin, a Philadephian business man and scientist, played a major role in drafting the Declaration of Independence (1776) and the Constitution of the United States. In the present era George Shultz has been an effective Secretary of State, after a career as a business school academic, and in construction with the international contractors, the Bechtel Corporation.

The leading figure in the foundation of Israel was Chaim Weizman, who already had an international reputation as a research chemist. Rajiv Gandhi has successfully assumed the position of Prime Minister of India after the assassination of his mother. He brought to an emotionally charged situation the professionalism of a trained aeronautical engineer and airline pilot. He has said that leading India is often a lot worse than flying through turbulence. 'There is sometimes tremendous slack in the controls, the inertial forces are incredible – and the passengers are very impatient'.

In Britain, the Prime Minister, the Rt Hon Margaret Thatcher P.C., F.R.S. M.A. B.Sc., worked as a research chemist, 1947–51, and was called to the Bar in 1954, specialising in taxation and patent cases. Sir Stafford Cripps also had a background in science and law. Before becoming prominent in politics he had a successful patent practice, including a major case over the rights to the patents on the pentode valve. In the years after the Second World War he set up the Anglo-American Productivity Council, through which the USA shared its business experience with Britain.

In Mrs Thatcher's second Parliament, out of 650 MPs elected in 1983, only 14 gave their occupations as chartered engineers or scientists.*

Occupation \ Party	Conservative	Labour	Liberal
Chartered Engineer	7	1	1
Scientific Research	3	2	0

2.6.5 Enhanced personal awareness

The person who can move relatively freely across the spectrum of disciplines and cultures is likely to be the one who is aware of their similarities and differences.

Conversely, the worst degree of ignorance is being unaware of one's fields of ignorance. If the framework of the body of knowledge and experience is recognised, then, with a little effort, expertise in depth in any area can be acquired. The computer 'menu' provides a model of how to dig deeper and deeper into a subject.

Busy professionals have limited time available in which to expand their spectrum of knowledge. The way that is best for one person, may not work for another – for example:

- through personal contacts, and by knowing the right people (the grapevine or old boy method)
- by feats of memory (the 'Mastermind' method)
- selective reading (for example, key abstracts, the *Financial Times*, and a limited range of journals and magazines)

---

* 'BUTLER, D. and CAVANAGH, D.: 'The British General Election, 1983' (London, Macmillan) p. 236.

Electronic publishing has not yet made a contribution to general awareness, but is invaluable for in-depth research via data bases. Most people will need the guidance of an information scientist, as, for example, through the service provided by the IEE Library. Quinn (1986) found that, although 60% of engineers in UK electrical engineering companies knew how to access data bases, only 5% used them often.

Awareness in breadth is a more personal intellectual process, involving conceptual thinking, pattern recognition, and a facility for seizing on analogies (which may turn out to be false). A practised mind will react intuitively, and the confident person will back his hunches, at least far enough to test an hypothesis.

Perhaps the most powerful analogy to apply is that of the feedback mechanism, and the application of systems analysis to all types of activity.

For the busy professional with limited time for personal study, there are technological aids. Broadcasts of Open University and other instructional material (audio or video) can be captured for review at one's convenience. 'Distance learning' updates the correspondence-course method with video tapes supplemented by coaching, and some very high-quality material is available (for example, from Henley–Brunel and from the Open University).

The IEE Professional Brief (1986) on 'Career Development' is most helpful, both with regard to objectives and to attitudes. It concludes with this quotation:

> The happiest and most successful person works all year long at what he would otherwise choose to do on his summer vacation (Mark Twain)

In industry and business, academic qualifications are no more than a launching pad for three or four decades of continuing education and self-development. Qualification, as has been said of money, is not everything, but it helps. Obtaining a specialist qualification or an M.B.A. serves as a discipline to complete the course of study, but this should not be at the expense of continuing awareness of the wider spectrum of contemporary technical and social trends and developments.

There is a continual stream of new books relating to management, some superficial, but both readable and good conversation pieces. The BIM publishes annually a 'top 20' list (Bibliography, Chapter 7.6), and also issues a series of 'reading lists' on aspects of management. The *Financial Times* and *The Times* include supplements on management books.

2.7 The nature of leadership

A natural leader will generally emerge in any group, particularly under the stimulus of a critical situation. But the nominal head of the group may not possess that capability: this is quite likely to happen in the high-technology industries, because people achieve initial promotion primarily through their technical expertise.

Innovative enterprises depend greatly upon effective team and project work; so an element in the design of any programme should be to ensure that key people possess the necessary personal qualities.

At any point in time, selection for an immediate requirement has to be from amongst those available. A far-sighted company will take steps in the longer term to develop these and other skills, so that appropriate people come forward as required. To do this effectively, there must be some understanding of the nature of leadership skills: on this subject much has been written. Jenkins (1947) made an analysis of 72 books and articles on leadership (and much more has been published since his study) – he found that few 'authorities' agreed upon the identity of the main characteristics.

Adair (1983) has quoted statements of considerable insight from the fields of sport (Mike Brearley, on the England *v* Australia cricket season of 1981), and mountaineering (Chris Bonnington, on leading expeditions). However, most analytical study, training and selection for leadership has been associated with the military and their staff colleges. What was written was autobiographical or by military historians.

Adair had experience both on the teaching staff of Sandhurst and later with industry. He is now Professor in Leadership at the University of Surrey. The insight from the Jenkins analysis was that the characteristics of leadership are specific to the particular situation – the conventional wisdom of 'horses for courses'. Adair developed the concept that there are three kinds of need always present:

- the need to achieve the defined task
- the need to maintain and motivate the team or group
- the needs specific to each person in the team

He further identified that these needs are interdependent, and diffuse into one another, so that they can be represented by three overlapping circles. It will be apparent to engineers that for any situation a polar diagram might be drawn, showing their relative weight and overall polarisation, which will be characteristic of the situation. This view rationalises the differences found by Jenkins in his analysis.

The value of this concept, for the better understanding of a complex phenomenon, is that subjective measures can now be applied to the influences of the three parameters. This immediately suggests what action is appropriate. The 'natural' leader is doing this subconsciously, but once the model has been defined, it can be taught.

Adair's quotations from Brearley and Bonnington show how a leader sets about involving individuals in the common task, and accepts their different personalities, and builds on them, as one would in selecting the cast for a play. There are many parallels between leadership, management and 'show business', particularly when the leader finds it timely to present and describe the mission and its objectives.

The emphasis in leadership training is how to apply these ideas and move away from merely discussing the needs, by introducing 'action-centred leadership'. This is the same idea as Revans (1971) 'action-centred management training'. In Adair's interpretation, as used by the Services, teams tackle physical tasks requiring ingenuity, organisation and vigorous activity, under expert observation. While originated for the military, such courses are now available to business and industrial organisations.

Similarly, the traditional 'Board Interview' method of selection was replaced or supplemented by weekend or three-day residential selection tests, so that people are assessed in action. This is an extension of the fact, well known to work-study practitioners, that a subject may modify the natural pace and style when initially under observation, but will revert to it if the observation period is prolonged.

There are many examples in business where the key to success or disaster has been the selection of the person to lead the enterprise, manage a project, or head and R & D team. The chief executive of an organisation needs the ability to select and develop further potential leaders. When large organisation structures are decentralised, as in the classic example of General Motors in the early 1920s (Sloan, 1963), or, since the 1970s, by companies such as ICI, Racal and GEC, the choice of the unit chief executive is crucial. As noted in Section 3.7, an effective central planning and review system will signal whether the leadership is effective, but if a replacement has to be made, there can be a setback measured in years.

Since Sloan, organisational style has moved through bureaucratic to highly consultative and participative, then back towards firm and positive leadership. These changes have been traced in recent books by Bass (1985), Lupton (1980), Margerison and McCann (1985) and Bennis and Nanus (1985).

2.8 Effective motivation of groups

The more fundamental differences in management styles and cultures were explored in Chapter 1.12. What may be seen to work well in Japan will have to be adapted even when an identical operating unit is set up in the UK, with British personnel. A subsidiary of a North American corporation will find it difficult to apply its standard procedures in a Latin American country, without some adaptation to the local culture. One of the strengths of UK companies engaged in international operations is that they can build on the earlier colonial traditions of accomodating the culture of the country.

A further difference that will influence management style is the stage of progress in education of the personnel available, and how familiar they are with the Anglo–American management conventions, which are by no means universally accepted elsewhere.

The 'loose-tight' style of management characteristic of many of the companies, classed as 'excellent', is particularly effective in the motivation of work groups. There is a formal planning and review process and, in the case of projects, a tight programme control system. But within these constraints a loose informal relationship is found, with the team leaders or unit managers spending perhaps half of their time mixing with the team, seeing and being seen. They delegate as much as possible, but are available to assist and coach, and are in a position to closely monitor results.

For the head of the unit to do this, he must manage his priorities and his time rather strictly. A good deal has been written about time management: Lakein (1985) and the BIM Checklist no. 1 and its reading list will be found helpful.

In a new unit, or a freshly formed project team, the leader has a unique opportunity to set the tone, and start as he means to go on. He has probably had previous experience as a leader or a member of projects, and is aware of some of the pitfalls, and how things could be done better.

In very large teams, as on a major development or construction project, a hierarchical organisation has to be designed. Burbridge (1984) has described this for a large power-station construction, and subsequent experience is described for the £1000 million project at Drax (1986). The NASA programme was considerably larger, and a recent review has been published by Beggs (1984).

Some of the key elements in building an effective team are:

- ensuring clarity in the statement and understanding of the mission, purpose and objectives of the project or task – if necessary clarify by further dialogue
- where there are clients, establishing an effective working interface, with the requirement specifications clearly stated and understood
- adequately briefing individuals joining the team; not being too busy to discover any special potential or capabilities
- in group sessions, identifying the several tasks to be performed, within the total programme
- where overlapping responsibilities come to light, resolving them by identifying the respective roles, for example:
  - responsible for execution
  - setting standards/defining specifications
  - need to know/liaison role
- detecting 'underlap' of responsibilities; things for which nobody is responsible
- laying down a pattern of regular review sessions on a calender of dates, and (in a complex project) at several levels
- setting up a project office, as a point of reference and informal contact

The open style of a project office or chart room offsets the intermittent character of teams that only meet together at a weekly or monthly occasion: the room is always there as an information centre, with updated data in chart form on display, preferably with a resident planning assistant or secretary. This is helpful where the team is mainly dispersed elsewhere, on site or in their parent discipline departments.

Both in Japan and in North America, 'togetherness' is stimulated with social events both at the workplace and at external venues. In the days of benevolent British proprietor management, annual outings were quite common, but now any paternalist flavour is best avoided: a more acceptable arrangement is for management to provide a budget, with the arrangements in the hands of an elected group.

The team leader or department head has a personal problem of maintaining an even-handed good relationship with all staff, but personally distanced from particularly close friendships. This is quite difficult when a promotion is made from within the team. It is easier if the leader comes as a sideways promotion from elsewhere in the organisation.

While reasonably tolerant of his subordinates, the head person has no alternative but to command their respect by setting an example, as a role model of high personal integrity and effort. The style of any new senior executive will rapidly be taken up by others, either for good or bad.

Freedom to lead a team in an imaginative and original way is possible within a large and somewhat bureaucratic organisation by adopting the formula of 'intrapreneurship'. This brings the entrepreneurial initiative within an established company, and a number of case histories are given by Pinchot (1985): it is a way of using 'high flyers' who otherwise might choose to seek opportunities elsewhere.

2.9 Industrial relations

Good industrial relations can be compared with preventive maintenance. Small progressive and informal businesses will have few problems if the management are able to maintain good human relationships, and are aware of their statutory obligations.

When organisations grow, the professionalism of a staff personnel function is necessary, so that company strategies and procedures may be adequately formalised.

A long-established manufacturing company will carry with it a history of staff agreements and attitudes towards Unions, which has to be taken into account when planning ahead.

New ventures in high technology usually treat all employees as qualifying for 'staff' conditions, which removes many potential areas of friction. The staff are aware of the progress and degree of success of the company, through their involvement, and an attitude of good communication. This tends to make unnecessary any formal negotiation of conditions. People experienced in the newer technologies are potentially in demand, and will tend to 'vote with their feet' as individuals, rather than bargain formally with management, as a group.

The danger point is when a new venture begins to mature, and the work force now feels that it needs Trade Union assistance in negotiating with management for its 'share of the cake'. This is a perfectly acceptable situation, if it comes about in an orderly way. But it can lead to serious deterioration of relationships if management tries to block the trend, and then later has to give in on Union recognition.

There are degrees of recognition: there is a tendency for some American-owned companies to frown upon any membership of Unions. A British company is more likely to accept that groups of employees will be represented.

The Union will generally seek full negotiating rights on terms and conditions of employment. In the traditional industries they have sought 'closed shop' agreements, and there have been recurrent difficulties with demarcation disputes between one Union and another.

Armstrong (1984 chapter 22) sets out a six-step management strategy for dealing with Union recognition issues. In arriving at its strategy a company should take an enlightened view of the positive contribution that a Union can make, both in terms of growth and of innovation or, alternatively, of an orderly restructuring and consolidation in a mature or diminishing market.

In the engineering and science-based industries there is both growth and consolidation, the latter in power plant and the heavier mechanical industries. The general picture of British industrial relations is focused on fairly static or declining industries, and the overall national growth rate is only some 3%. The innovative industries may have growth rates of the order of 15% or more, and are really a different kind of animal.

The easiest pattern of growth is when each activity can be set up on a 'green field' site, unrelated to the back history of a long established business that has seen many changes. Both new units of British companies, and those of Japanese origin setting up in UK, have made satisfactory agreements for a single Union to represent all skills on a particular site.

This greatly simplifies the traditional procedures. New industries present new opportunities and attractive working conditions: to such initiatives, Unions generally react in a very positive way.

Large construction sites have been the focal points for special difficulties, which were the subject of two Reports in 1969 and 1970 (Burbridge, 1984). Since then, great improvement has been achieved by the development of project strategies for the co-ordinated management of all aspects of a construction site and its many contractors; e.g. with the successful completion of the Drax power station (Drax, 1986) and other large projects. Emphasis is now laid on establishing clear lines of communication on the site, high safety standards, single-status canteens and transport facilities, and a good standard of temporary living accomodation and social facilities.

Guidelines to good practice in industrial relations are included in BIM's series of Management Checklists, mentioned in Chapter 2.5.3. The Code of Practice of the Advisory Conciliation and Arbitrations Service (ACAS) is helpful, together with the publications of the Industrial Society, and the Institute of Personnel Management.

ACAS is responsible for the 'Industrial Relations Handbook', published by HM Stationery Office (1980). This outlines the historical background of the Trade Unions, and of collective bargaining, and the arrangements followed in various industries. Comparative studies of practices in different countries are published by the International Labour Office, Geneva (which also has a London address).

It can occur that professional engineers are expected to belong to a Trade Union, and may then or later have managerial responsibilities. Notes on this situation are contained in the IEE's Professional Brief, 'Guide to Trade Union Membership'. There are useful books by Cooper, and by Whincup.

2.10 Effective communication

2.10.1 Putting ideas across

It is often remarked that engineers communicate and express themselves poorly – this not only limits their effectiveness in the technical sphere, but can seriously prejudice their success as managers.

Sir Peter Walters, who began his BP career in the supply and development department when he was 24, rose to be Chairman of the company (Britain's biggest) and, in 1986, president of the Institute of Directors, He has said that he first came to the attention of senior management through the ability to write a good report.

A specialist may do brilliant original work in isolation, but if it is to have any practical application and benefit to the community, the ideas must be spread. They will be recognised and adopted in direct proportion to the clarity with which they are expressed.

The advice of Benjamin Franklin (1706–90), a successful scientist and businessman in Philadelphia, turned statesman, was:

> . . . before you sit down to write on any subject, you would spend some days in considering it, putting down. . . every thought which occurs to you. . . Examine them carefully. . . to find which is to be presented first to the mind of the reader. . . that he may understand it and be better disposed to receive what you intend for the second. . . (and so on). . . each preparing the mind for that which is to follow.

To communicate effectively we must adopt a 'systems approach'. Today, being aware of this concept, we have the tool to explain more succinctly than couid Franklin. We know how to write a broad specification and expand in detail in terms of the subsystems. One of the big intellectual inventions was the 'black box': you worry only about its input and output; designing the insides is a separate task.

Having identified the idea we wish to put across, we must match it to the intended audience, rather than enjoy the conceit of a highbrow exposition. It implies no disdain of your audience's capabilities to consciously 'tune-in to their wavelength' – we are only following the principle that for maximum transfer of energy we must match impedances.

The process of interpersonal communication needs to be planned and 'engineered' as thoroughly as any other task of systems design. We are fortunate today to have available a whole range of media denied to the classical writers and orators, particularly the ability to augment oral with visual communication, if only at the level of the flip chart.

2.10.2 Varieties of communication

The following check list of aspects of communication illustrates the wide range of coverage:

Writing: end-uses

- requirement specifications: logically structured, as a basis for tenders and contracts
- project proposals: to secure approval at higher levels
- technical writing
  - summaries of discussions or investigations

- reports on progress
- technical and instruction manuals

Presentation
- verbal explanation of a proposal or a report, to a higher level
- instruction and training
- major audio-visual development of theme for wider audiences

Leading a discussion or a meeting
- free, unstructured discussion (T-Group method)
- informal discussion or training session, with summing-up
- formal meeting, with Chairman, agenda and minutes
- formal negotiation, commercial or trade union

Public relations
- presentations to external audiences
- conference presentation, and prior organisation
- relations with the public media: technical Press, general Press and television

It is an instructive exercise to take a particular subject or theme, and then consider what adaptation is involved in putting it forward effectively in each of these different ways.

A realistic example would arise from the decision to plan and build a new factory, office or laboratory. There are the early stages of deciding what is wanted and involving those who will use the facilities. There is the contractual side, arrangements with Local Authorities and utilities, local environmental matters, informing the public in the locality, and general prestige publicity in the company's market sector. If the whole is not planned systematically on project lines, things will go wrong, not only mistakes but because, for some aspects, nobody had been designated as responsible.

Communication is of course a two-way process, with outputs and feedback. Any proposal to do things differently, that involves people, will generate reactions, and some one should be listening for them. The proposals may have to evolve in an iterative way, with emphasis on the positive features.

2.10.3 Advice on communication

What Franklin said has been written about by many others, and these BIM Checklists (with suggested reading) are helpful:

List no	
13	Planning a meeting
32	House journals
50	Giving a talk
61	Using visual aids effectively
66	What is public relations?
76	Effective communications

In addition to the books mentioned in the above checklists, the following are of interest (see Bibliography, Chapter 7.2):

Armstrong (1984 chap 24)	Fletcher (1983)
Arthur (1984)	Irvine (1971)
Duerr (1971)	Jefferies (1983)
	Scott (1985)

These IEE Professional Briefs are helpful when writing for publication:

Technical Report Writing
Symbols and Abbreviations
Technical Writing for Publication

The subject is treated in depth in the *IEEE Transactions on Professional Communications* (USA, published quarterly).

2.10.4 The art of speaking

Some individuals are natural public speakers, but, for most, this facility comes from effort and practice. It is easiest to work up gradually, from round-the-table discussion with acquaintances to larger audiences.

Management development programmes usually include group sessions, where members take it in turn to sum up the conclusions from a task, and make a brief presentation.

The effort of preparing visual aids is well worthwhile, having a threefold advantage:

- the ideas or points to be put across must be clarified into the concise form that will fit onto a flip chart or projector transparency (and still be visible)
- communication is reinforced by using the two senses, sight and sound, and handouts can be made available as notes of the talk
- it is much easier for the presenter, because the audio-visual material serves as the prompt notes, so the speaker can be more spontaneous, and concentrate attention on the audience

When a talk seems to be extemporary, and has plenty of eye contact with the audience, it is much more acceptable than direct reading from a script. A degree of expression through body language does help, with some reaction to the audience such as a change of pace if there is a murmur of assent or of amusement. Jay (1970) is helpful, from his experience with BBC TV, since applied in the production of training videos.

2.10.5 Staging a discussion

When a meeting is primarily to stimulate discussion, it needs to be firmly stage-managed by the leader or chairperson. Those present will respond better to the informality of being seated in a circle, rather than in rows as for a lecture.

An effective approach is to structure the introduction to include a mention by each member of his affiliation – name cards help. After a period of free discussion, the leader should move naturally into a feedback review round the circle, so each can speak and relate what went before to his particular experience.

This does ensure that everyone contributes at least once. At this point the leader can sum up. Then, if appropriate, a further aspect can be introduced for discussion. In a continuing programme, those present should accept personal tasks, either for reporting back at a later session, or to be summarised on paper. The whole object is to encourage active participation.

2.10.6 Formal negotiations

These meetings require careful preparation, by identifying all likely options and eventualities. Some initial 'staff work' may be appropriate – preliminary talks about talks, so that the main event starts with an agenda already agreed.

The participants at a formal meeting will be there to represent their company or organisation: they need prior briefing, and must be clear about their policy objectives and limits of their delegated authority. This applies equally, whether the negotiations are with Trade Unions or on commercial matters.

Negotiation with customers or suppliers ought not to be approached in a adversarial manner, but in the spirit of seeking an agreed basis that is mutually advantageous. Either side, by adjusting its technical or logistical requests, may be able to optimise cost and value.

A thorough analysis of the options should be made before a negotiating meeting, and a general strategy agreed. This will identify points at which alternative proposals or concessions might be put forward. The senior delegate must secure clear terms of reference on the limits of the commitments he may make on behalf of his organisation.

Unexpected and cost-effective ideas may arise. The story is told of how Henry Ford specified to close limits the details of the packing cases in which components were to be delivered: the boards of these cases were then ready to use as the flooring of the Model T Ford, a car that sold for $100 in the 1920s.

2.10.7 Public Relations

Public relations are a sensitive area in most organisations; the large ones usually have available an experienced PR manager. A unit manager within a large Group will have the benefit of this advice when it is needed. He probably also has standing instructions on how to react if approached directly by the press.

In independent companies it should be generally known who will act as spokesman in response to any enquiries from the media.

Both positive and negative situations may arise. A branch unit of a large company, or an independent enterprise, both need to secure and maintain goodwill in their local community. This requires positive initiative from time to time, in order to be known to the Local Government representatives and officials the head of local services, the constituency's MP, and the local Press.

Goodwill from these initiatives will pay off if any negative situation develops, where the company may be subject to critcism or objections to aspects of its activities.

All organisations should have a pre-agreed 'disaster plan', whether or not they operate dangerous processes: communications and PR in an emergency should be part of this plan.

2.11 Reading list

ACAS (1980): 'Industrial relations handbook' (London, HMSO) 354 pp.

ADAIR, J. (1983): 'Effective leadership – a self-development manual' (Aldershot, Gower/Pan) 228 pp.

ARMSTRONG, M. (1984): 'Handbook of personnel management practice' (London, Kogan Paul, 2nd edn.), 416 pp.

BURBRIDGE, R.N.G. (1984): 'Some art, some science and a lot of feedback' *IEE Proc.*, **131**, Pt. A, pp. 24–37.

CLUTTERBUCK, D. (Ed.) (1985): 'New patterns of work' (Aldershot, Gower) 160 pp.

HARVEY-JONES, J. (1988): 'Making it happen – reflections on leadership' (Glasgow: William Collins), 266 pp.

LOCK, D. and FARROW, N. (Eds.) (1983): 'The Gower handbook of management' (Aldershot, Gower, 2nd edn.), 1200 pp.

PEACH, L.H. (1983): 'Employee relations in IBM' *Employee Relations*, **5**, (3), p. 1720.

PINCHOT, G. (1985): 'Intrapreneuring' (New York, Harper and Row) 236 pp.

SCOTT, W. (1984): 'Communication for professional engineers' (London, Thomas Telford), 240 pp.

TOMLINSON, H. (1983): 'Productive labour' *IEE Proc.*, **130**, Pt. A, p. 1930.

'Opportunities in management for professional engineers' (1988) (London, Peter Peregrinus) – 21 articles by IEE contributors.

Chapter 3

Managing financial resources

3.1. Summary

A brief appreciation is provided of the history, evolution and segmentation of the profession of accountancy. This indicates that much change has been going on, paralleling what has happened in the engineering profession.

The general principle in the UK (and rather less so in the USA) is that the professions such as medicine, law, finance and engineering should be self-regulating under their Charters, with additional mandatory guidelines provided by legislation, from time to time.

Engineers have the Engineering Council; Lloyds and the Stock Exchange are expected to put their own affairs in order, and the accountants have to satisfy the requirements of the Companies Acts, most recently those of 1981 and 1985. That of 1981 for the first time introduced an obligation to follow principles, rules and standards which, as agreed under the terms of the Treaty of Rome (Fourth Directive), achieve harmonisation of practices within the European Economic Community.

An engineer, scientist or technologist who becomes one of the team managing an enterprise must gain an insight and appreciation of the contribution of his colleagues in the functions of accounting and finance. To adequately follow what they say and do, he must have at least a superficial understanding of their language and terminology.

If he is to succeed with technical proposals he must know how to make a convincing presentation in terms of cost effectiveness. The accountants' world should not be a mystery to the engineer (and vice versa), for they share the linguistic ability to express what they mean, in terms of numbers.

Before investing too much of his personal time in a study of the financial aspects of management, a non-accountant should home-in on and clarify his own objectives.

Does he want to:

- start up as an entrepreneur in his own modest business venture
- have a general understanding of the accounting function, in a medium-sized company

- appreciate how the function works, in a large Group, when he is in a Division, or a subsidiary company

This book can give no more than an introduction to concepts and general principles, together with simple illustrative examples. It does provide a trail into the specialised literature to guide the reader in pursuit of particular objectives and needs. Chapter 7.3 contains an extensive bibliography, including general reading, grouped to correspond with the Sections of the text in this Chapter 3.

To cite just one of the books, John Sizer's 'An insight into management accounting', is particularly helpful, and each chapter contains references for further study. First published in 1969, it has been reprinted 14 times (up to 1985) and, as a Penguin, is inexpensive.

A selection of the standard books on the subject will be found in most Public Libraries, as they are used by people preparing for the professional examinations in accounting and banking.

In recent years financial management has been included in the lecture programmes and in the published *Proceedings* of the Institutions of Electrical and Mechanical Engineers: some references are included here, and they are useful because this is material prepared by and for engineers.

The mystique of accountancy, for those outside the profession, has been dissipated with the advent of integrated computer-based management systems. The 'systems approach', whether in engineering or in management, provides a rationale for cutting across the traditional departmental and functional barriers, while retaining clear ground rules and delineation of where responsibilities lie, and who does what.

3.2 Accounting, engineering and auditing roles

Charles Babbage (1792–1871), best known as a mathematician and scientist, and for introducing program-control into computing, was also a good instrument engineer and able to make detailed manufacturing cost analysis of the components of his machines as illustrated by Urwick (1951). His weakness was in strategic planning, for his projects were never completed, owing to overruns in both cost and time.

This was a common problem among the early masters of engineering, who made every decision themselves, including their own financial arrangements. To mention just a few:

James Brindley	1716–1772	canals
James Watt	1736–1819	steam engine and governor
George Stephenson	1781–1848	steam locomotives
Isambard Brunel	1806–1859	railways, bridges, ships
William Henley	1813–1882	electrical instruments and cables
Thomas Edison	1847–1931	professional inventor

Henley's case history by Anderson (1985) is a cautionary tale of rapid growth in the new technology of the day, but failure to watch cashflow and to delegate responsibility, and to adapt to the growth of the business. Bamford (1986) and Cattermole (1987) record similar aspects of the history of instrument manufacturers.

Somehow, in the latter part of the 19th century, a dichotomy arose between the people who made things and the people who looked after finance, to the extent that in some companies until very recently engineers and accountants were hardly on speaking terms. Even if they were, they did not understand each other's jargon. They had little appreciation of each other's contributions to the success of their enterprise. Happily the position has improved, and there has been convergence, reaching its fulfilment in what we know as 'systems engineering'.

In major contemporary projects, such as power-station construction, the Thames Barrier, NASA space projects and the Channel Tunnel, it is essential that the several functions and contributors work together as a team. There must be enough mutual understanding to ensure effective and coherent communication at the professional and commercial interfaces. We have also seen the language barrier overcome by the project teams of Concorde, the Airbus and the Channel Tunnel.

This applies equally, but in simpler terms, in more modest projects of construction and product development. We have come full circle from Babbage's experience. It is computer systems and management-information data bases that now ensure commonality in the work of the several functions of an enterprise.

The external auditing role of the public-practice accountant is a statutory requirement under the Companies Acts. Most larger organisations now also make their own internal self-audits of their operations.

This is a demonstration of a concept that Peter Drucker (1955, Chapter 11) describes as 'management by self-control'. Here each person is responsible for checking his own performance or the results of his department, within a framework of objectives and goals, agreed at the next level of management.

This self-auditing principle has come a long way from the Victorian view that no one can be trusted, and must be visited and reported on at intervals by the Inspecting Officer. This attitude goes back at least to Roman and Greek times, and we can read that Samuel Pepys, as Secretary of the Navy, made an annual inspection visit to Chatham Dockyard in the 17th century.

Direct personal responsibility is the concept underlying present-day quality management systems. In British Standard BS5750 on 'Quality Systems', basis of a draft International Standard ISO 9000, we have a demonstration of an approach to self-audit that applies to every function in a company, and to personnel at every level.

It is likely in 1986 that financial audit regulations will be liberalised, by excusing the large number of smaller companies from the statutory requirement, and placing the responsibility directly on themselves, as a contribution towards reducing the burden of their administration.

3.3. How accountancy has evolved

It is useful to the engineer to have some background understanding of the accountant's profession. In the beginning there were counting and token systems used for trade, and this takes us back to Babylonia and Mespotamia, about 3000 BC. At this time, in the area that is now Syria and Turkey, the calender was invented, so there was the basis for primitive annual accounts some 5000 years ago. The chart in Fig. 3.1 shows the three streams that followed:

- data processing, from the abacus to 4th and 5th generation computers
- number systems, geometry, calculus and Boolean algebra
- bookkeeping, leading to financial and management accounting.

Recognition of the importance of numerical systems is typified by these two quotations:

> Count what is countable, measure what is measurable and what is not measurable, make measurable
>
> Galileo, 16th century

> When you can measure what you are speaking about and express it in numbers, you know something about it: but when you cannot measure it, when you cannot express it in numbers, your knowledge is of a very meagre and unsatisfactory kind
>
> Lord Kelvin, 1883

The Greeks, Egyptians and Romans all had the ability to measure accurately, and to lay out buildings with precision. However, we may not appreciate how inconvenient to manipulate were some of the early systems which employed various bases other than decimal and binary, and with difficult number characters (as in the Roman numerals).

Decimal notation appeared with the Hindus, about 500 AD, and the Chinese had the first mechanical computing aid, the abacus, about 1100 AD, and it is still in general use both to add and to multiply. Pascal's adding maching in the mid 17th century led directly to the mechanical and electromechanical calculating and bookkeeping machines of the early part of the 20th century, such as the Brunsviga, Comptometer and Burroughs.

In the 17th and 18th centuries there was great interest in 'automata', such as clocks with moving figures, and mechanically played musical instruments. The skills existed to create mechanical bookkeeping, but the possibilities had not been recognised or the demand had not been created. Jacquard punched-card control of looms dates from 1728, but the process of 'technology transfer' took until 1890 when Hollerith applied the principle, initially for collation of the United States census returns.

Colossus (at Bletchley) and ENIAC (at Moore School, Philadelphia) were the first electronic computers, in the period 1944–46, and the earliest business opera-

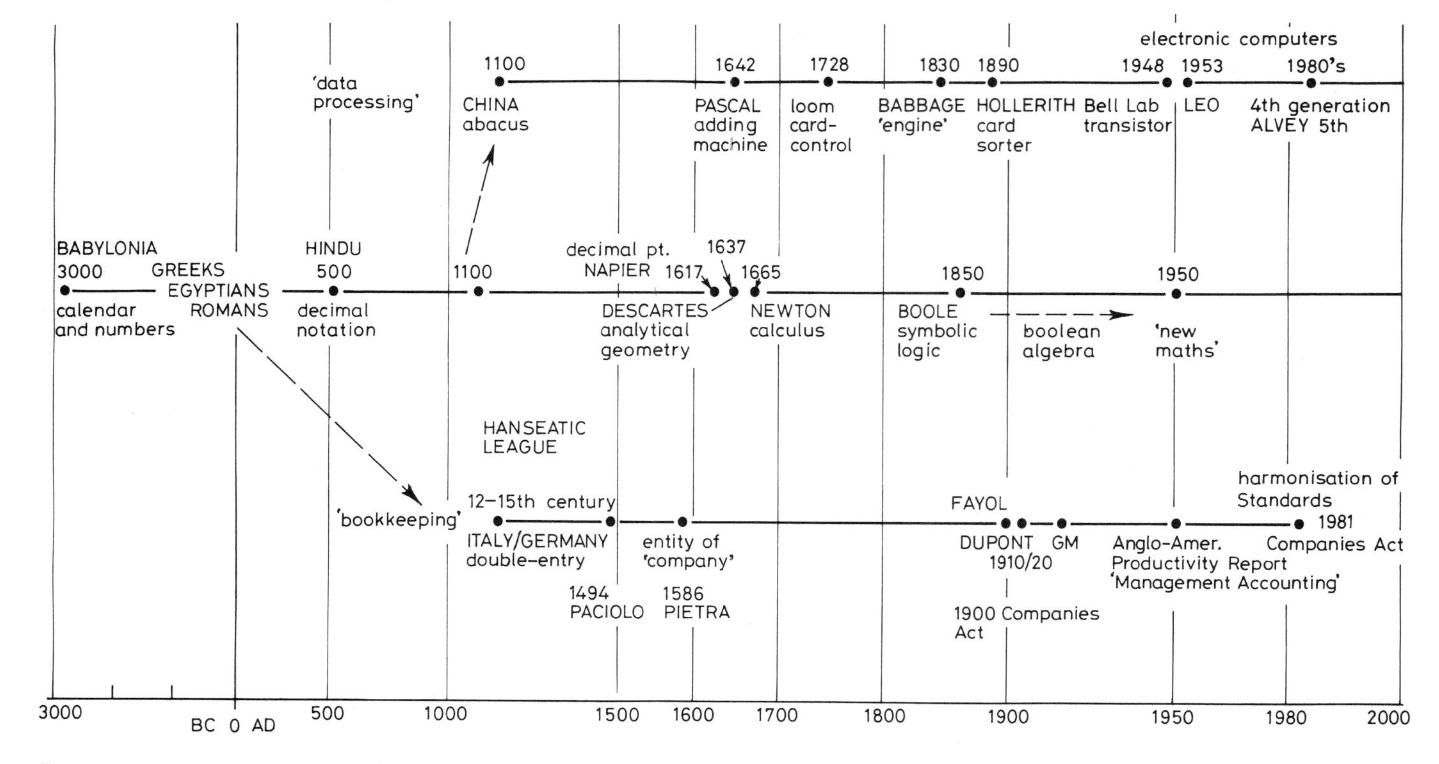

Fig. 3.1 *Historical development of the practice of accountancy*

tion was probably the LEO machine, carrying out in 1953 the payroll and the daily teashop orders for the J. Lyons Company, remarkable in that they developed the machines themselves from pioneer work started in 1947 at Cambridge University. There followed the Elliott 401 and the IBM 701, also in 1953. The British Computer Society was formed in 1956.

These machines used thermionic valves and generated much heat, and it was the invention of the transistor in 1948 at the Bell Telephone Laboratories that led to the next phase of powerful high-speed computers. With the development of integrated circuits in the early 1970's, the powerful mini- and microcomputer became possible, and in the 1980s these have been linked as systems in local area networks.

The development of 'bookkeeping', has an interesting history, as the big breakthrough was the introduction of the double-entry system during the 12/15th centuries, but it was not described publicly until an Italian, Luca Paciolo, published an instruction manual in 1494: 'Summa de arithmetica geometria proportioni et proportionalita'. There is a translation by Brown and Johnston (1963).

This is widely accepted as the foundation stone of modern accounting and bookkeeping systems, and it is thought that the method was in use much earlier, as the first known double-entry books were in Genoa in 1340, and were already in a highly developed form.

Records of contracts have been found in the ruins of Babylon, and accounts for farms and estates were kept in ancient Greece and Rome. The Italian, German and other merchants of the Hanseatic League, who developed the double-entry system, probably kept their methods secret, as was the custom of the time. The significance to a trader sending expeditions that might not return for months or even years, was that he was able to control separately his assets and his liabilities or commitments, without falling into the error of supposing that cash in hand is a measure of the health of the business.

Another conceptual development that followed was when Don Pietra, a Benedictine monk, postulated in 1586 that a business enterprise should be regarded as an economic entity, distinct from the personal fortunes of its owners. The next century saw the beginning of the growth of the great trading companies which were to open up India, the Far East and Canada.

The Venetians, as a maritime people, were among the first to develop trade with the East, but lost their position in the 16th century, because they had not the technology of the bigger ocean-going ships necessary to voyage West. The initiative then passed to the Portugese and Spanish, and later to the English and Dutch traders.

In the 19th century, the emphasis in business was literally on 'keeping books' at the clerical level. It was said that the main role of accountants was to sort out the affairs of bankrupts, and this unfortunate state is often mentioned in Dicken's stories. The importance of controlling cash flow was emphasised by Mr Micawber's comment:

> Annual income twenty pounds, annual expenditure nineteen nineteen six, result happiness. Annual income twenty pounds, annual expenditure twenty pounds nought and six, result misery.

The limited liability Company came into being with the Act of 1900, and in recent times the requirements have been updated in the Companies Acts of 1948, 1967, 1981 and 1985. The two latter bring into effect the harmonisation requirements of the EEC, which are spelt out in a series of Directives. The Fourth Directive of 1976 requires a consistent form of presentation of Annual Accounts throughout the EEC.

While preparing for the necessary changes in practices, the accountancy profession had also to adjust to the very high rate of inflation during the late 1960s and 1970s, and high and variable interest rates.

As in engineering, there has been much activity in the development of a series of International Accounting Standards (IAS), and further development of International Standards is in progress, as a means of harmonising the presentation of the affairs of multinational companies.

Similarly in engineering, the British Standards Institution has taken an active lead in developing the methodology and format of Standards that are proving to be models for international acceptance.

3.4 Financial and management accounting

As in engineering, there are distinct branches of the accountancy profession, and six major bodies.

Institute of Chartered Accountants in England and Wales
Institute of Chartered Accountants in Scotland
Institute of Chartered Accountants in Ireland
Chartered Association of Certified Accountants
Chartered Institute of Management Accountants (previously ICMA)
Chartered Institute of Public Finance and Accounting

There is a joint Accounting Standards Committee, formed in 1969, representing all six bodies. The ICMA originally represented 'cost and works' accountants, and has broadened its scope in recent years. Of particular interest to engineers are the courses run in the Diploma scheme of the CACA, designed for the non-accountant.

There are three main streams in professional accountancy:

- public practice: covering company accounts, audits, insolvency and as consultants on taxation
- as Directors and employees in commerce and industry
- in the public sector: local government and the Civil Service

Within public practice, individuals usually specialise on a particular aspect, especially in the large firms. Accountants who work as business executives tend to develop an all-round view of their immediate environment, and often prove to be the best qualified candidates for the role of chief executive.

As a simplification, it can be said that, in business and industry, the practice of accounting historically divides into the two areas;

Financial accounting

Preparation of the annual accounts in the form required by the Companies Acts, and carrying out Audits.

A *small client company* may use a firm of outside Accountants for both functions, or employ a qualified accountant for work other than external audit.

In a *medium sized firm* the Company Secretary frequently combines the function of accountancy with the responsibility of dealing with all legal matters.

Management accounting

This function provides an operational service to management which now, with computer assistance, can include real-time cost information on a same-day basis. This is a more effective management control than formal annual or quarterly Accounts: the service provided will include assistance to functional managers in developing their budgets, and in making the financial justification for new projects and plant.

The Annual Accounts are a statutory obligation, but may not be ready for publication until some 18 months after the start of the reference period. It is now rare for companies not to have a separate management-information system. The analogy has been made that, depending on the historic information of the annual accounts is like driving a car while looking out of the back window.

'Cost accounting' originated in manufacturing departments during the early 1900s. Even with clerical methods and punched cards many companies were able to report a week's operating results on the Monday morning following. This emphasis on rapid reporting originated in the United States, and developed into a complete system of budgets and feedback of actual performance, thus becoming known as 'management accounting'.

Standard targets of output were set, corresponding to 'standard costs', and any variance reported was watched closely. Thus the role of the management accountant becomes that of:

> the presentation of accounting information in such a way as to assist management in the creation of policy and in the day-to-day operation of an undertaking.

The responsibility for acting on this information lies with the appropriate operational manager; so, for the whole organisation, lines of responsibility must be clearly defined.

It was Fayol (1908) in France who pioneered clear functional definitions in management from the 1890s. His work was not widely known until the publication of his book 'Administration industrielle et générale' in 1918, followed by an edition in English in 1929.

A comprehensive financial budgetary information and control system was first developed on a large scale by the DuPont Company in USA in the 1910s, and extended through their assocation with General Motors in the early 1920s, and rapidly became the normal pattern of management control in the large corporations. As staff migrated to smaller companies, the system spread and was taught in the American Business Schools, quickly achieving widespread adoption.

The United Kingdom, and indeed Europe as a whole, had generally remained with historic financial procedures until after the Second World War. In 1950 a British management accounting team made a classic report on what they had seen in the USA, on a visit arranged by the Anglo-American Council on Productivity.

The Report of the Team included the statements:

> ... the greatest single factor in American industrial supremacy is the effectiveness of its management at all levels ... one of the measures of this is the ability to use the means of production at its disposal – both men and machines, to produce the greatest output at the least cost ... this is one of the main differences between American and British practice in management

Considerable progress has been made since the above was written, and now an integrated management information system can remove most of the tedium of clerical analysis, and do it accurately, and very much more quickly, using a central database. For example, the following can all be linked in a common system with access through terminals at virtually every working point:

- material and parts list of a product (possibly as the output of a computer-aided design system)
- incoming orders for products
- advice of firm delivery date
- material and component stock commitment and control
- purchase orders, for stock replenishment
- production scheduling and machine loading
- goods inwards and rejection documentation
- despatch documents
- customer invoices, accounts and credit notes
- general management returns and statistics on performance
- feedback to the quality management system feedback
- scrap analysis
- stock reconciliation (with discrepancies signalled)

As all data is related to a single database, transcription errors are eliminated. The management accounting data is also more readily reconciled with the financial accounting statements, as they are based on the same sources of information.

Standard software packages are now available, providing these facilities for straightforward applications such as manufacture in light engineering, and for the design and assembly of electronic equipment. In particular situations the standard software may have to be 'customised', and 'fourth generation' systems enable this to be done in plain language by the user's staff.

3.5 Familiarisation with the conventions and practices of accounting

Anyone intent on serious study should consider taking the management qualification in accounting provided by the Diploma of the Chartered Association of Certified Accountants. Some 80 Polytechnics and Colleges offer part-time courses, mostly evening, covering one academic year. There are also available 'distance learning' courses, and video presentation from several sources, and it is desirable to check what courses are running in one's immediate locality.

A more superficial approach to familiarisation is to study one or two of the introductory books mentioned in the Bibliography, and then continue with the more specialised ones appropriate to your particular needs. Bookshops and Public Libraries stock only a selection of all the current texts; so it is best to see what is immediately available. The paperback editions are particularly good value.

Books on accounting present their subject in rather general terms, applicable to any business. In applying the concepts it must be remembered that engineering businesses can be untypical, with their special characteristics of rapid innovation and obsolescence, and high complexity and unit value of the products.

Some introductory books make their presentation in the form of a narrative about a fictional firm: a good example of this is the BBC's 'Hardy Heating Co. Ltd', now out of print, but available in some libraries.

It is the custom in this field to present extensive examples of forms of account, and to practise with 'question and answer' books. As mentioned in Chapter 3.1, the engineer familiarising himself with accountancy should be clear about his objectives. If he is joining a small company in a senior capacity, or starting his own venture, he should concentrate on the literature on small businesses, including the legal obligations, especially if he is to be a Director. He can benefit from membership of the Institute of Directors, which provides a number of publications on the responsibilities of company directors.

If it is his own business, it is wise to seek out a local accountancy firm of progressive outlook, on whom he can depend initially for guidance. In the stages of growth of a new venture there will be an appropriate time to add to the management term a professional accountant, either a keen young one or, possibly, a person of well rounded experience recently retired from a go-ahead company. The right person will know how to introduce a comprehensive computer-based management accounting system.

In a medium-sized or large company an engineer's need will not be so much to take the initiative, but to have a good understanding of what his professional accounting colleagues are doing, and how their systems work. In a large Group which is divisionalised or operates with many relatively small subsidiaries, the professional engineer may have the responsibility of General Manager or Managing Director.

There is then an acute need to be aware of the financial and legal requirements, but guidance will be available from Group Headquarters. He will have to work with a system of periodic controls and reviews peculiar to that Group and, if he takes a job elsewhere, may have to adjust to a change of style of management and a rather different system of reporting and presentation.

3.6 The main financial accounting statements

3.6.1 Introduction

There is a legal requirement that a formal Balance Sheet be published annually, after independent audit; it states what a business enterprise:

owns (its assets)
and what it *owes* (its liabilities)

at a particular date. The other two principal statements are the 'Profit and Loss Account' and 'Sources and Applications of Funds'. The main components of these three are:

Profit and loss account
- Turnover
- Trading profit
- Interest on borrowing
- Tax on profit
- Net profit after tax
- Dividends to shareholders
- Net profit transferred to reserves

Balance sheet
- Fixed assets
- Current assets
- Creditors
- Net current assets
- Net current liabilities
- Total assets less current liabilities (= net assets)
- Capital and reserves

Sources and application of funds
- Funds generated from operations
  - Trading profit
  - Interest
  - Depreciation
- Net changes in stocks, debtors and creditors
- Net proceeds from sale of properties and other assets

Total funds generated

Application of funds

Change in net borrowings

According to the nature of the business these headings will be expanded in detail, and supported by 'notes to the accounts'. Engineers will notice that, although accountants now rarely use black and red ink, as the distinction between profit and (loss) and credit and debit, it is the custom to use brackets rather than the mathematical negative sign.

3.6.2 Presentation of Accounts

Since the 1981 and 1985 Companies Acts, and under the influence of the EEC, presentation now conforms to a set of Standards, and has become more uniform and more readily comparable, but will differ in detail between the several types of business, the main categories being

- service businesses
- trading businesses
- manufacturing businesses

The distinction is now blurred owing to amalgamation of diverse businesses into Groups. Any well run Group will make the distinction in its internal operating accounts, but these are not usually available to the public. Companies in the USA have always tended to be willing to disclose more details of their operations.

The main public companies will send copies of their Annual Accounts and Chairman's Report on request, at the time the details are announced in the Press, and it is useful to secure a cross-section of these as examples of presentation of the data.

3.6.3 Bookeeping

Because classical double-entry bookeeping goes back some five centuries, the terminology sits rather oddly in a contemporary computerised system.

There are two types of 'books', which literally were leather-bound volumes, at one time written up meticulously with quill pens. The advent of mechanical accounting machines in the early 1920s (for example, by Burroughs) introduced with it loose-leaf systems: the machines were a combination of typewriter and mechanical calculator, plus a number of mechanical storage registers.

The types of books are:

Journals: a daily listing of transactions, sales purchases and credits
Ledgers: the records of personal accounts with individual customers, entered up from the daily journal records, and nominal accounts of expenses etc.

Engineers will recognise the possibility of linking these two parameters as a matrix, so that an entry is only made once, and then totals taken horizontally and vertically. This is a facility available today on a microcomputer spreadsheet.

The tedious and meticulous work of clerical entry and cross-entry is known as 'journalising'. For every transaction there had to be a voucher, such as cash slips, check books, stock cards, deposit slips, time cards, sales slips, sales invoices, credit memos and purchase invoices.

These were sorted, analysed and entered into the journal as the first formal record of a transaction, the completeness of the vouchers being checked against their printed serial numbering. In a large business there would be separate journal volumes for different types of transactions.

The second stage of recording was posting in the ledger, the process of making a cross-entry from each item in the journal to the appropriate account in the ledger. The bound ledger was laid out in double-page spread with debits on one side and credits on the other.

If there are no errors in a day's postings, the total of all debits must equal the total of all credits, and testing this is called the 'trial balance'.

The possibilities for clerical error are considerable, not only omissions and numerical errors, but also errors in sorting and classification. Alterations of the inked entries were frowned upon, and had to be formally initialed, and the incentive to get it right was that the bookeeper worked late until the trial balance was proved.

Subsequently, at audit time, much attention would be given to sampling and checking the vouchers to establish if they corresponded with the book entries, and this tedious task was the first stage in apprenticeship to the accountancy profession. There are spectacular advantages from logical use of computer coding and processing of transactions and posting to the 'books', but the audit branch of the profession is now faced with the much more difficult task of evolving adequate procedures for audit of information processed and held in computers. With clerical methods and serially numbered vouchers there was a ready 'audit trail' that could be followed, and an equivalent means of cross-checking is needed when computers hold the data.

At intervals, a Financial Statement is prepared by summarising the trial balances. Traditionally this was monthly, but it may be weekly or for a four-weekly financial period.

Some local accounting firms will provide a 'package' service to small businesses whereby they undertake all bookeeping on a computerised basis, deal with the payroll, and work up the formal accounts. Self-employed craftsmen quite commonly put all their papers in a box, which they hand to an accountant or an experienced bookeeper to sort out and put in a form acceptable to Customs and Excise (for VAT), the Tax Office, and Companies House (if a registered company).

A company in new technologies is likely to have rapid growth, and if this is forseen, it will be important to have good formal accounts from the start. The reason for this is that, when external financing is required, the records will be examined with an expert eye, and the availability of a clear record of 3–5 years of progressive growth in turnover and profit will be highly regarded. If the first two or three years are poorly presented, this much credibility time is lost.

3.6.4 Finance as a process flowsheet

Just as we have translated from the language of the accountant by describing the journal/ledger relationship as a matrix and spreadsheet, the whole financial system in an enterprise can be presented as a process, by means of a flowsheet. Fig. 3.2 is a simplified example, showing the components of the three main statements of account.

The professional economists are conscious of the analogy when they speak of 'the velocity of money', as the flow rates differ in the several parts of the flowchart. For example, if a bakery business pays out wages on Thursday, and an employee takes his family to McDonalds that evening, the bakery could be delivering more Big Mac buns on Friday, and so show a credit back in their own accounts for that same week.

In contrast to the 'fast-food' business, money put in a savings bank or an insurance may not be called upon for a matter of years. Money invested as the permanent capital of a business is tied up until that business is liquidated: however, there is a market for selling or buying the shares if the company is quoted on the Stock Exchange, or dealings can be made between individual buyers and sellers.

The financial community distinguishes between these different kinds of money and the different rates at which it can circulate in the national economy. The definitions are:

- M1 = 'money' in the narrow sense of banknotes, coins, 'current accounts' or anything immediately cashable, like Travellers Cheques (known as 'retail M1', at the personal level)
- M2 = M1 + savings and short-term deposits that receive interest
- M3 = M2 + large denomination interest earning deposits (known as 'domestic or sterling M3', when holdings of foreign currencies are omitted)
- M4 = M3 + private sector holdings in Tax Instruments, Treasury and commercial Bills
- M5 = M4 + Building Society deposits

Although these national indicators are of a rather specialised nature, they are significant to larger businesses as signals of the current and future trends in national expenditure and in the cost of money.

The flowsheet highlights the three kinds of money employed in a business, distinguished by their different 'time constants':

- the permanent capital, put up by the shareholders, who are the owners
- money borrowed for periods of medium term (of the order of 5 years) which must be repaid when due
- short-term arrangements for which the limit is weeks or months, and in general less than a year

An expanding business will experience problems of shortage of cash, because it has incurred expenditure on materials, labour and expenses for several months before it can collect back cash from customers who buy the additional production.

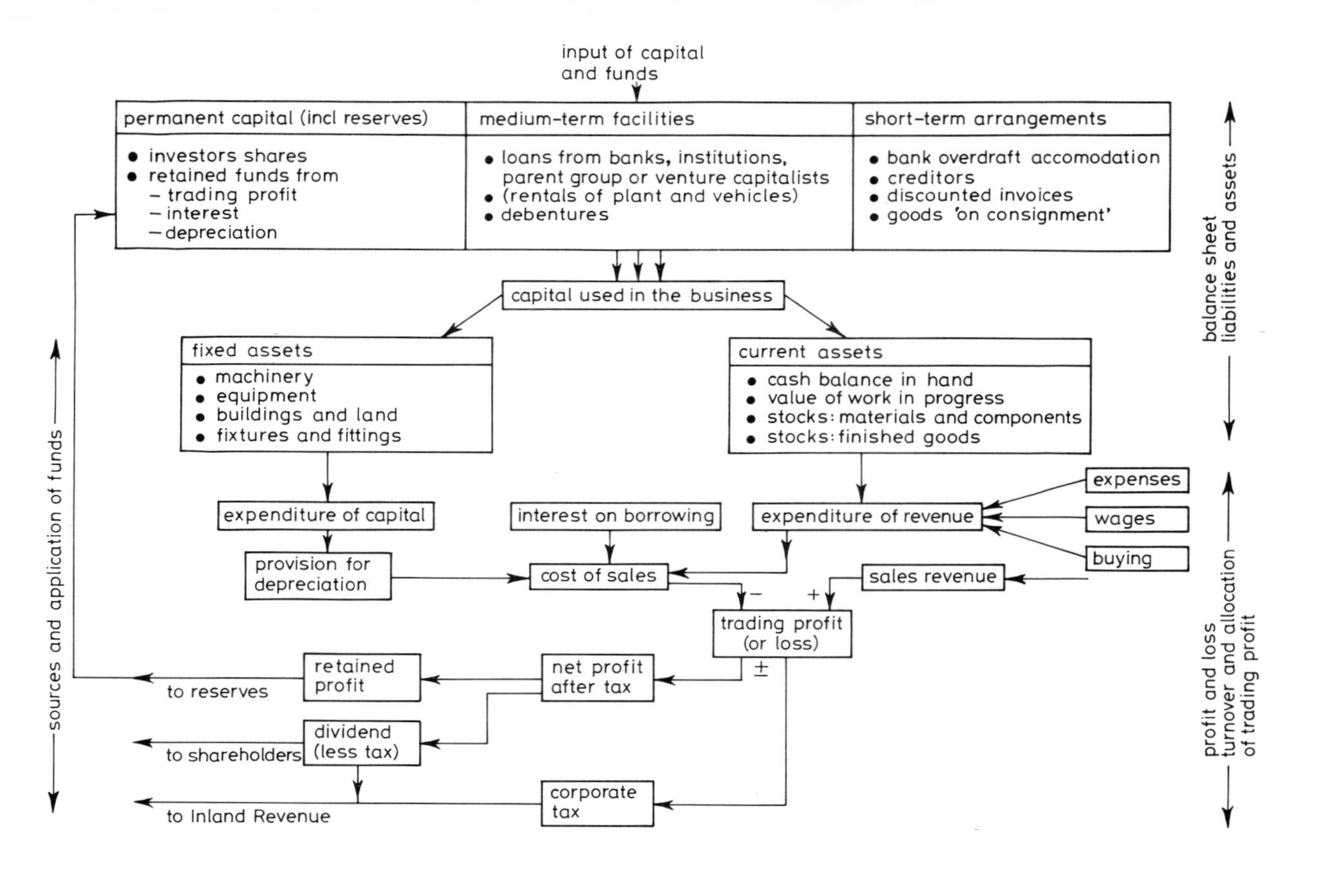

Fig. 3.2 *Finance as a process flowsheet*

In these circumstances there is available 'off balance sheet' finance, but this will not be granted to 'lame ducks' which cannot show a healthy order book and profit record.

The three main sources are:

- obtaining new plant and vehicles via rental agreements from leasing companies (the major Banks operate leasing subsidiaries)
- factoring: obtaining an advance against invoices, so that money is received immediately goods are despatched; the advance may be of the order of 60–75% of the invoice, in return for a service charge of $1\frac{1}{2}$–2%
- subcontracting work out, as an alternative to spending capital on the extension of the company's own facilities (e.g. for metal presswork or finishing, or plastic moulding)

It is reasonable for an enterprise to use these financial services, provided that the business is healthy and has a good quality of management. Factors, leasing companies and subcontractors are all in business to make profits for themselves. A company should review its use of such facilities regularly to assess whether or not it could be obtaining these profits within its own operations.

3.6.5 Key financial ratios

For a true understanding of the health of a company, it is necessary to know what arrangements or commitments have been made. Company accounts now contain Notes to the Accounts, and full disclosure is required when a company's equity is placed on the Stock Exchange: a prospectus, as published in newspapers, may occupy several pages of small type.

When such detail is not readily available, a powerful method of analysis is to use Business Ratios. There are two main applications of the method:

Internally: changes in the year-to-year trend in the pattern of the business can be indicated, and acted upon.
Externally: a company needs to know its competitive position within an industry or market sector, and ratios of different elements of cost can highlight weaknesses and strengths.

An example of the need for internal ratios was a company designing and making mechanical handling equipment. It was steadily becoming unprofitable, although it was advanced technically in design and development. The reason was that it was buying an increased proportion of hydraulic and electronic control components, where in the past it had itself manufactured the mechanical equivalents:

$$\text{the ratio} \quad \frac{\text{value added}}{\text{turnover}} \quad \text{had dropped}$$

and this trend and its consequences had not been spotted by the Board of Directors.

Facilities for external comparisons are provided by the Centre for Interfirm

Comparison, which process the confidential data provided by members, and then makes available to them comparative data for a particular business sector, that does not identify individual firms.

Alternatively, a company can make comparisons for itself by using data obtained by market research, both from published trade statistics, and by sample surveys. The use of 'management ratios' is explained by Westwick (1973).

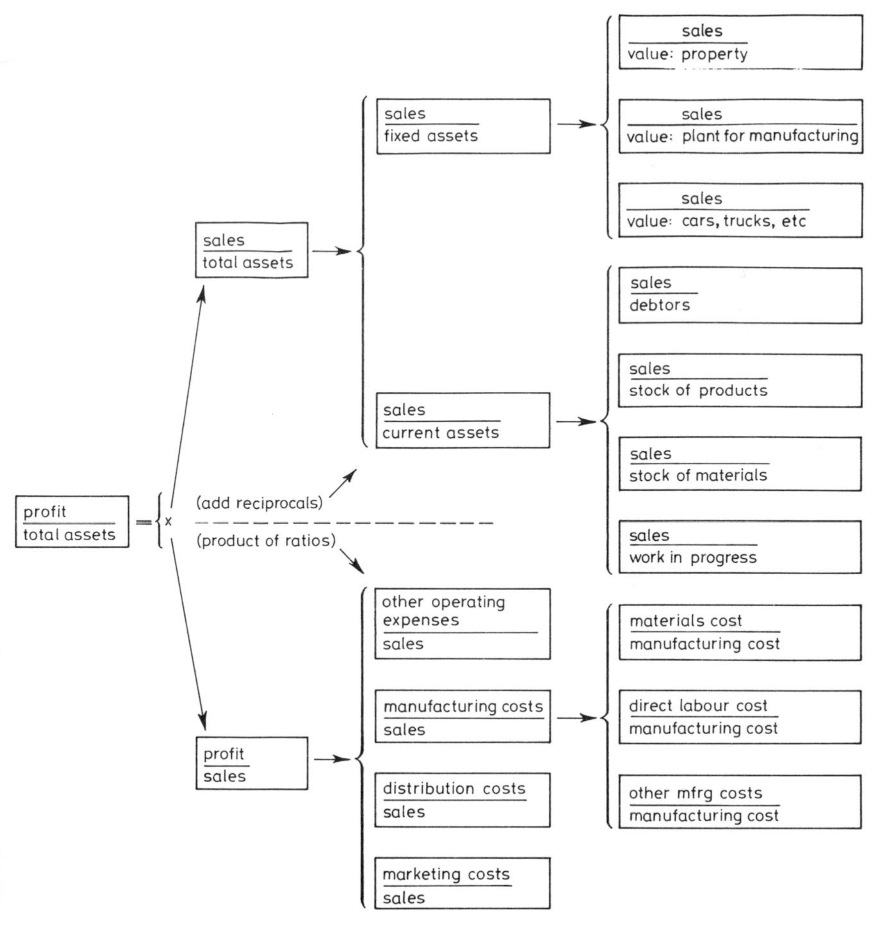

Fig. 3.3 *Key financial ratios: expansion in detail*

High accuracy of analysis is not necessary to signal whether a company is losing its place competitively: the most common error is to assume that all is well because there are moderate annual increases in turnover, when both the ratios of share of market and of profit/turnover are steadily falling.

A simple principle applied in the analysis of ratios, is that they be factorised: $A/B = (C/B) \times (A/C)$

$$\text{thus:} \quad \frac{\text{profit}}{\text{capital employed}} = \frac{\text{sales}}{\text{capital employed}} \times \frac{\text{profit}}{\text{sales}}$$

This example is important to firms that may make a large investment in computer-aided design and manufacture: the important ratio then is the sales in relation to capital employed, and the implication is that the new and expensive equipment must be operated for more hours per week, to earn its keep, which implies multi-shift working on the manufacturing side, supported by at least some duty staff on the design side to resolve any queries that arise.

An expansion of the factorising of the ratios is shown in Fig. 3.3:

- where the denominator is common in the factorised ratios, their overall ratio is the product of the secondary ratios
- where the numerator is common, the reciprocal of the overall ratio is the sum of the reciprocals of the secondary ratios

This concept of differential analysis can be applied to every aspect in the operations of a company, and the implications of these are examined in considerable detail in management-accounting textbooks.

3.6.6 Treatment of R & D expenditure

R & D expenditure is covered by the Statement of Standard Accounting Practice (SSAP) No 13, published in 1977. The 1981 Companies Act does not require special treatment of R & D, but full disclosure is called for in the EEC Fourth Directive, which the UK must follow in due course. The International Accounting Standard (IAS No 9) is virtually the same as SSAP No 13.

Three broad categories of research and development are defined:

(*a*) pure or basic: original investigation for new knowledge
(*b*) applied research: original investigation directed towards a specific aim or objective
(*c*) development: the use of science or technical knowledge in order to produce new or improved products, processes, systems or services, prior to the commencement of commercial production.

These definitions are useful when preparing budgets, as some firms use the term 'R&D' rather loosely also to include straightforward engineering design and drawing-office expenses. The SSAP definitions leave the boundaries of 'engineering' unclear, and also the interfaces between engineering design and development and production engineering. The latter function is of increasing importance in process engineering and automated processes of production.

R&D budgets vary very much between industries, being 2% of the UK's overall GNP, but less than 1% of sales in mechanical engineering, and over 15% in aerospace and electronics (no doubt due to taking the wider interpretation of R&D).

SSAP 13 requires that both pure and applied R&D expenditure be written off in the year incurred: there is the exception that, for a clearly identified and defined development project, the costs may be carried forward, being treated as an investment. Clearly described stages of a project are necessary, in order that the value of what has been done may be audited.

The engineering industry stimulated changes from earlier traditional accounting practices. The misfortunes of the original Rolls Royce Company, prior to its bankruptcy in 1971, arose because, for a decade or so, part of the R&D expenditure was shown in the accounts as an asset, which allowed a profit to be declared. Then it was necessary to abandon the carbon-fibre blading of the otherwise successful RB 211 engine development, and large sums were written off from these book assets. The Government came to the company's assistance, and it was reconstructed.

In November 1970, *The Engineer* made the editorial comment that this had 'demonstrated once and for all that engineers on their own are incapable of running large companies'. The journal also commented that the signs were there in the accounts from previous years, but no one was prepared to ask what was happening. Again, in June 1971 this journal said that 'nine times out of ten the appearance of a lame duck is a reflection on management and its failure to come to grips with the situation'.

Another instance in the late 1960s was the take-over negotiaton between Pergamon (publishers) and Leasco (computer leasing), where the value of assets was in dispute: what is a fair price for a used computer, late model mainframe?

There were also difficulties in that same period when GEC wished to absorb AEI: the Directors of AEI forcast profits for the year of £10 m., but subsequently the outturn was a loss of £4·5 m.

The fact that these problems arose in high-technology industries is no coincidence: traditional accountancy was based on an expectation that one year would be much like another, with no fundamental underlying changes, such as now occur in the more dynamic market sectors.

When R&D expenditure is carried forward to later years, it is said to be 'deferred', with reference to its appearance in the accounts for the current year. Special conventions apply to large manufactured units, such aircraft or generator sets, which are sold in limited quantities. Then the deferred expenditure may be written off as each unit is sold. In this connection 'sold' means completed and invoiced, not the more general usage of a firm order booked.

With large units the difficulty is always present of estimating how many will be sold, and the write-off per unit may be based on optimism, and not be realistic.

Another category of R&D expenditure is the construction of fixed assets or facilities, that may be used over a long period. The expenditure will be capitalised and written off over the estimated useful life, or the life of a specific project. When attempting to justify such major facilities it is first necessary to establish the company's policy, and also to take advice on the treatment for tax purposes.

3.6.7 Treatment of Stocks

The valuation of stocks and work in progress is covered by the Standard SSAP No 9, which originated in 1975. In any particular year the value placed on stock can have a considerable influence on the declared profit. For example, the operating profit from a turnover of £10 m. might be assessed as 10% (£1 m), with material stocks £2 m and work in progress £2 m. If the potential error in true value of the total of these two items is ± £1 m, we either have twice the profit, or no profit.

The value of work in progress can become negligible if a project is aborted, and materials and components can physically deteriorate, become obsolete, or be subject to market influences. A notable example are integrated circuits and, in particular, memory chips, where the replacement value dropped by a large factor during periods of oversupply, as in 1985. The more advanced the technology, the greater the risk (Fig. 1.12).

Companies that consume the more expensive raw materials in large quantities have to deal on the commodities markets, and will tend to make long-term contracts for metals such as copper and aluminium, and if these extend beyond one year, they will have special treatment in the accounts for taxation purposes.

External auditors experience some difficulty in assessing the true value of high-technology stocks, and will put on record the criteria they have employed. A company must decide which way it will treat the turnover of stock, either:

	first-in, first-out	(FIFO)
or:	last-in, first-out	(LIFO)

The one or the other is the more conservative, dependent on whether prices are falling or rising. When prices are rising, LIFO charges current costs to current production, while retaining a reserve stock bought cheaply. FIFO may be essential where materials have a finite life (for example, PCBs and wire-ended electronic components tend to tarnish, increasing the incidence of soldering faults and reducing reliability). An incidental advantage of the 'just in time' method of integrating production scheduling and stock control is that the risk of deterioration is removed.

Most companies operate a formal internal audit, which includes the stock-taking function. Continuous (or cycle) stocktaking is the process of taking stock, section by section, against a calender programme. Now that most records are carried on computers, it is essential that the computer transactions cease during the period that a particular block of stock is being physically counted.

A special category is 'goods on consignment', which means that they are in the custody of those other than the owners. so they need to be clearly identified as such. The same applies if they are 'free issue', from a customer for incorporation in his order.

A small manufacturer should balance the risks of stock holdings against the alternative of carrying no free stock and buying at a higher price, as required, from a distributer. Both methods can be used together: for example, a manufacturer of electronic equipment, making up in batches of ten, may order only the exact requirement of electronic components from their sources, so that he has no residual stocks: any shortages or replacements will then be obtained as required from a local distributer, most of whom offer same-day service.

During periods of inflation it is advisable also to present accounts on a 'current-cost' basis, valuing assets at replacement cost rather than at historic cost, which throws up a less optimistic level of profits: the accounting conventions are set out in the Standard SSAP 16.

3.6.8 Corporate taxation

A company will usually rely heavily on the advice of its external accountants in taxation matters. Treatment of various allowances will vary from year to year at Budget time, as the Government attempts to influence the trends in the business economy through the tax system.

There is an extensive literature on taxation accounting, and the leading firms produce interpretive guides for their clients very quickly after the Chancellor of the Exchequer has made his March Budget statement.

Assessable profits for tax purposes are not the same as the net profit shown in the published accounts. For example, the Tax Inspector ignores the depreciation that the company sets against profit, substituting for it the current capital allowances.

A company will usually take expert professional advice on the options to minimise tax, that are legally acceptable. As the Tax Year runs from 6th April to 5th April following, company accounts for two calendar years have to be used as the basis for the figures of one tax year.

Other responsibilities are placed on a company, which has to collect and account for:

- Income Tax from the salaries of employees, using the PAYE tax tables
- Income Tax due on dividends paid to shareholders, which is shown as a 'tax credit' on the statements they receive
- Value Added Tax (VAT) due to Customs and Excise: the net amount of VAT added to sales invoices (outputs), less the amounts for VAT that the company pays to suppliers (inputs): this account is made up and settled quarterly

In consequence of the period of high inflation during the 1970s, 'current cost accounting' was introduced as a mechanism for presenting a realistic picture of profitability, allowing fortuitous gains due to increased value of stocks, and potential losses as a consequence of replacing stocks and plant at inflated prices. The Accounting Standard SSAP 16 (1980) defines the procedures. A current-cost balance sheet may accompany and supplement the accounts based on historic costs. Considerable extra work is involved, and some companies have discontinued the current cost statement in the years that inflation has been low.

This alternative presentation does not affect the liability for tax, but it may provide evidence in the event of negotation over the Tax Assessment. It is a useful tool for the formulation of company policy, as it highlights the effect on the currently declared profits of:

(–) replacing material stocks, at currently enhanced prices
(–) replacing essential plant, from a 'current cost' depreciation reserve fund
(+) products sold at current prices, but manufactured at historic costs

With modern stock-control and progress systems, it is good management practice to maximise stock-turn rate, and minimise work in progress. This reduces the effect of differences between current and historic costs.

The unfavourable circumstances that can get a company into serious difficulties include:

- holding a relatively large stock of finished machines and spares, which are now known to be technologically obsolete, with minimal market demand
- when design and development costs have been shown in the accounts as an asset, but the projects or products have proved to be abortive
- when specialised process or production plant is shown as an asset, but has become obsolete or uneconomical

From tax considerations it is best to 'bite the bullet' and write off such stock and other assets in the accounts for the year. This will either drastically reduce the profit, and hence tax payable, or show a 'tax loss' that can be applied to adjust tax liability in other years.

It may seem odd, but agreed 'tax losses' becomes assets when a company that has performed badly is valued in takeovers.

The knowledge that finished stock is virtually unmarketable may exist in a company, amongst the technical or sales personnel, but not be apparent to accountants. The auditors should detect a serious situation if they call for sales figures and relate them to the stock on hand. It is their responsibility to certify that the accounts 'give a true and fair view of the state of affairs of the company'. Their audit of necessity represents only a sampling of everything that might be checked, and unconscious or conscious overvaluation of stock and other assets may go undetected for a time.

There are tax advantages for a company by obtaining a proportion of its working capital as long-term loans from financial institutions, rather than by increasing the investment from the shareholders. These institutional investors will wish to see a healthy ratio of 'times covered' (gross income over interest due on their long-term loans).

A Board of Directors may be faced with the dilemma that the tax advantages of writing down the value of doubtful assets is in conflict with their wish to maintain a favourable level of declared gross income.

A fine line of distinction may exist between prudent stewardship and moral obligation. When companies are found to have run into financial difficulties, in retrospect it can be seen that the seeds of misfortune were present several years earlier. It may be questioned whether those responsible were fools or knaves, and whether their financial advisors could reasonably have been expected to identify the unfavourable trends. The first public signal of trouble is often the 'qualification' of the accounts by the auditors, and notes of disclaimer such as 'stock as valued by officers of the company'.

There are two contrasting ways of managing tax affairs in smaller or medium sized businesses controlled by a family, or a founder and his associates:

- a family-owned business with fairly static prospects may be run so as to maximise the personal pursuits of the proprietors, with a pair of Porsches, and other perquisites, but at the expense of declared profitability

- an entrepreneurial company with continuing growth prospects is better run on a tight budget of expense and overhead, so it builds a record of profit over several years: it can then be sold or become a quoted company, making the founders millionaires; they will have chosen to take long-term capital appreciation rather than short-term income.

Some notes on taxation and the individual, either when self-employed or working abroad, appear in Chapter 5.6.

3.7 Aspects of management accounting

3.7.1 Budgets and Controls

Business Planning, and its quantification as financial planning, should be carried out formally and systematically. A typical approach is to plan in detail for the next financial year, and in outline for 3–5 years ahead. The longer-range plan is then reconsidered annually and rolled forward a further year. There will usually be an associated profit plan and forecast.

The way to avoid escalating problems (such as the RB 211 story outlined in Section 3.6.6) is to follow a discipline of relating technical programmes to company budgets. These should be put together with the agreement of all key people at the several levels of management.

The feedback is then completed by making the head of each function personally responsible for reporting any significant deviation in his part of the programme, so that prompt corrective action is taken.

For this to work, there must be clear definitions of who is responsible for what at each level in the management hierarchy, and for each function and subfunction, taking into account that:

- overlap of responsibility is divisive, and can lead to confusion and misunderstanding
- underlap can leave critical areas unmonitored

There may be difficulty with the perceived meaning of words. An engineer will think of the verb 'to control', and expect to find an active and complete closed-loop system. Traditionally an accountant used 'control' as a noun, in the passive sense of a standard of comparison for checking against budget.

To be fair to the profession, accountants are now very aware of the concept of feedback control, and the need for this feedback to be rapid. They may be less perceptive about instability in feedback systems, and consequent undershoot and overshoot in relation to targets.

Concepts of management that are relevant, which engineers will recognise without further detailed explanation, include:

- management by exception
- hierarchy of control

- frequency of cycle of review
- time span of discretion

A small working group, with open relationships, can get along very well with informal and infrequent contacts. In a large organisation, with its hierarchy of levels of supervision, management and direction, a systematic approach is needed. Programmes and budgets will be promulgated downwards. Conversely, each level should communicate upwards significant exceptions or variance, taking corrective action or seeking advice on what action to take. If this style of 'management by exception' is understood and accepted, senior management can assume that 'no news is good news'; they have the back-up protection that cross-checks are implicit in the periodic financial progress statements.

This will work if an open culture is established where people will speak up whenever necessary. The opposite situation has been called the 'Chinese syndrome', as it is said that the bearer of bad news to the Emperor could expect to be beheaded. A more effective climate is where he is disciplined if he does *not* promptly pass the bad news to his superiors.

The counterpart to 'exception reporting' is to carry out a regular cycle of reviews. In any organisation with several levels, these reviews should be linked, so that

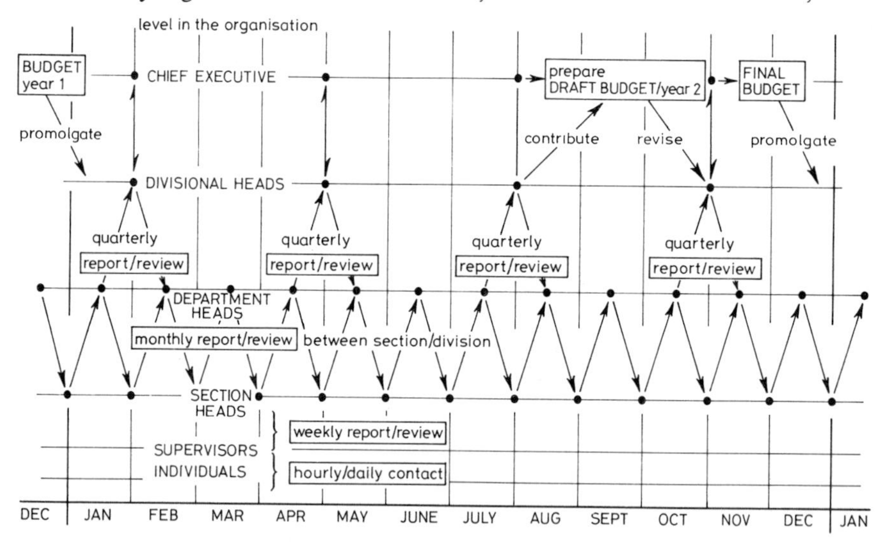

Fig. 3.4 *Budgets and controls: pattern of reporting results and reviewing action*

upwards and downwards communication proceeds in a regular synchronous manner. Dates of review meetings should be inviolate, so that those concerned can plan their diaries round them.

The chart in Fig. 3.4 illustrates that the pattern of reporting results, and reviewing action. The sketch shows that department heads have a monthly review with their section heads, the data input being:

- costs input from the finance function
- technical progress reports
- interactive data from the other departments

Following a monthly review, the section head should communicate downwards whatever is necessary to his supervisors, and through them to individuals.

The sketch indicates a quarterly formal review by department heads with the chief executive. One would also expect to find continuous informal contact, and *ad hoc* meetings on specific issues as they arise. Towards the end of a financial year, all levels will contribute upwards towards the formulation of the programme and budget for the following year.

The 'time span of discretion' is a helpful concept that originated in the work of Wilfred Brown and Elliot Jaques (Brown and Jaques, 1965, p 240). It is defined as 'the maximum time-lapse over which a person is required to exercise discretion in his work without that discretion being reviewed'.

In the model illustrated, supervisors report and review with their section heads weekly, and they have daily or hourly contact with individuals under their control. In this example, a division head (or the MD of a subsidiary) is assumed competent to run his sector of the business for a period of three months between formal reviews. At Chairman level the span may be 3–5 years, which is the time it may take to change the style of the enterprise.

Particular businesses vary in style, and those that have operated a very 'tight' system of review tend to become more relaxed as their senior managers become accustomed to a disciplined approach. Harold Geneen describes in his biography (Geneen, 1985) the system he operated in ITT in the period 1960–78, when very detailed reviews of each operating unit were made monthly, backed by voluminous centralised financial controls: it was said that senior executives spent most of each month preparing for and attending these meetings. Geneen believed that General Managers of subsidiary companies would learn from each other's experiences, and usually 120–150 of them attended the monthly meetings.

Turner (1985) has recorded Lord Weinstock's approach to the control of some 200 businesses that make up GEC. Attention is focused on a single monthly sheet of financial results and key ratios, but planning is largely left to the Heads of units.

A Managing Director's budget will usually be accepted, the critical attention being directed at its achievement. Weinstock is quoted as saying 'if a proposition is well presented, it usually takes me 20 minutes to give an answer'. Turner records that both the documentation and the decision times were far shorter than had been encountered elsewhere. It has taken time within GEC to appoint and train managers who understand the company's particular ideology and style.

It is feasible for the chief executive of a large company to use a computer terminal to access any operating data, for any depatment, and the vendors of systems may have encouraged the use of this feature, because it is possible to do so.

Although available in an emergency, the facility should not be abused by those at the top, because the consequences will be:

- they overload themselves with detail
- formal reviews are short-circuited
- the responsibility of middle management is reduced by over-riding their authority
- operational management should be signalling the exceptions upwards, together with their proposals for corrective action

A firm's annual budget is properly seen as a model of what may happen, worked out in detail. It will contain various assumptions for the sales of different products and services, and for the productivity levels, labour and material costs. It is now feasible to represent the parameters in the model by equations that evaluate the influence of changes in other factors. This can be done by spread-sheet methods, so that the effect on the budget of market forces can be evaluated quickly. Also, trial calculations can be made of the consequences of alternative strategies or tactics.

£'000

OUTPUTS → / INPUTS ↓	BUDGET ANALYSIS PER PRODUCT					
	A	B	C	D	E	totals
COST OF RESOURCES EMPLOYED						
materials	136	216	630	410	227	1620
direct labour	87	98	305	205	116	811
indirect labour*	40	32	130	65	63	330
technical labour	22	15	50	43	30	160
accomodation	30	32	140	58	50	310
administration*	8	7	28	14	13	70
other o/h*	14	11	48	24	23	120
sales expense	43	39	180	27	31	320
other mktg expense	25	31	110	50	44	260
packaging	18	13	52	45	42	170
transport	14	16	60	23	17	130
capital charges*	73	59	237	126	104	599
cost of sales	510	570	1970	1090	760	4900
total sales	700	550	2300	1150	1100	5800
gross margin	190	−20	330	60	340	900
margin/sales %	27.1	− 3.6	14.3	5.2	30.9	15.5%

Fig. 3.5 ***Operating budget and profit plan for a medium-technology company manufacturing five main product groups, and with about 100 employees***

This presentation is a 'spread sheet' or 'input-output' matrix: it can be updated period by period through the year as the actual figures become available, and as marketing tactics are varied: it can provide an updated forecast of the year's profit outturn

*Totals allocated in proportion to turnover: other costs are actual

The example of a spread-sheet (Fig. 3.5), representing the annual budget of a medium-sized company, shows five major product lines. If there is the opportunity of extra business, the effect on the several elements of cost may not be linear, as additional resources have to be sought elsewhere (for example, subcontracting), or overtime worked at a premium rate. On the other hand, the amount of 'fixed' overheads will be adjusted very little, and will be spread over a larger turnover.

In this 'matrix' form the consequences of change in the budget can be revised very readily during the course of the year, and alternative tactics can be tested before decisions are made. The types of variables that could be fed in might include:

- cost of a component is £Y, but £$(Y + Z)$ when subcontracted
- distribution costs are £x/day/vehicle + £y/mile, divided by numbers of deliveries
- key materials can be brought cheaper if volume increases

Williams (1985) describes examples of how market planning strategies can be evaluated using a microcomputer, and provides specimen software programs.

3.7.2 Future value of money

It must be recognised that all forms of financial planning and budgeting are no more than models, and the farther ahead they project, the less probable is their fulfilment.

There have been periods in history when the value of the pound and the rates of interest were almost constant, e.g. 1919–39. The experience of recent years is that both factors can vary greatly, increasingly under international influences. Interest rates have settled at a relatively high level, and a good deal is at stake in taking financial commitments beyond a year or so ahead.

In compensation, there are ways of hedging the risk, by putting money into fixed-interest securities, and by dealing in commodity and money futures. For example in a relatively stable period, $US can be bought for delivery in 12 months time for a premium of about 4%, or about 1½% for delivery in 3 months. Such facilities are important to a company manufacturing and exporting large plant, or engaging in major overseas contracts.

These uncertainties are superimposed on the basic calculation of future value. The present value of a future expectation of income or liabilities is computed as 'discounted cash flow' (DCF).

Thus if an account for £1000 is not payable until 12 months hence, the payee (creditor) loses the use of that money for a year, while the payer (debtor) has the benefit of it. In practice, the interest he can earn by depositing it is less than it would cost the payee to borrow an equal sum meantime: this is because the banks themselves may have an operating margin of some 4% or so between their lending and deposit rates.

Thus it pays a company group to operate a central 'treasury' as a service to the subsidiaries, absorbing their short-time surplus cash balances, and providing deposit and lending facilities. The picture is further complicated because liability for tax on interest received and tax allowances enter into the calculations.

In the simplest cases of cash flow, where I = annual interest rate (as decimal) and N is the number of years:

accumulation factor is $(1 + I)^{+N}$, and
discount factor is $(1 + I)^{-N}$

If the cost of borrowing for a year the £1000 of delayed payment is 15%, then the initial present value is only £850 if the interest is paid in advance, or £870 if it is paid in instalments through the year. Most accounting textbooks deal with 'discounted cash flow' (DCF) calculations, and Bunn (1982) gives a good introduction.

Relatively high interest rates are likely to continue, and are linked to rate of inflation. Longer-term projects can only be justified financially if the rate of return is high enough to cover:

- interest earning on capital
- recovery of capital over x years
- a risk factor, for
  - technical failure
  - error in market assessment
  - vigorous competition
  - early product obsolescence

For example, if a manufacturing process takes £1 m. and one year to build and commission, using money borrowed at 15%, the outlay at the end of year 1 is £1·15 m.; then the present value of target levels of profits is as shown in Fig. 3.6

Typically, it would be unwise to count on returns beyond year 6, as the technology may be obsolete, or the demand and price may have become unfavourable.

The break-even point in the example of Fig. 3.6 is when the profit per year is £343 000, for recovery of the investment in real terms by the end of the five year

operating profit per year		£0.2 m	0.3 m	0.4 m	0.5 m	disc. factor
discounted profit at end year:	2	0.174	0.261	0.348	0.435	0.870
	3	0.151	0.226	0.302	0.378	0.756
	4	0.132	0.197	0.263	0.329	0.658
	5	0.114	0.172	0.229	0.286	0.572
	6	0.099	0.149	0.199	0.248	0.497
total years	2–6	0.670	1.005	1.341	1.676	3.353
initial outlay		1.150	1.150	1.150	1.150	
present value from 5 year operations		– 0.480 loss	– 0.145 loss	+ 0.191 profit	+ 0.526 profit	x £m

Fig. 3.6 *Present values of future profits from operations over the next five years*

life. But the company has also tied up £1·15 m. of capital that otherwise could have earned interest at no risk. If past experience shows that, for every five successful ventures, there is one that is a write-off of the investment, then a further element of £1·15 m./5 must be recovered from the 5-year earnings; thus:

break-even operating profit	£343 000
risk-factor allowance	£230 000
minimum target profit	£573 000

This example is, of course, simplified, as it omits the important factor of normal tax allowances, and special allowances for new enterprises, and special regional development grants.

The normal provision for depreciation would be:

transport equipment:	25%	on written-down value
general fixtures on furniture:	10%	on written-down value
plant and machinery:	15%	on written-down value
freehold buildings:	2%	on cost
freehold land:	nil allowed	

In accounting for the value of assets, there are three distinct sorts of costs:

Original cost: this is the cost when acquired, and is recorded as 'book value'
Replacement cost: this is the current cost of an equivalent replacement, and will tend to be more than the original cost, by the amount of inflation, but a lot more if the appropriate replacement is equipment of higher technology.
Opportunity cost: this is the value of an asset, for another or alternative purpose; the current cost of occupancy of a property may be relatively low (for example, as a warehouse), but if that site could be let for another purpose at a relatively higher rent, the 'opportunity cost', is the net cost of not realising this potential.

A company may hesitate to invest £1 m. in a new plant or process to satisfy its existing market better, more competitively or more profitably. But if it fails to take the initiative while competitors do, it is likely to lose most of its share of that market sector, and that side of its business will become worthless.

A company as a profitable going concern has a marketable value, especially with a Stock Exchange listing. But if it ceases to have a good profit record, the value drops dramatically, and may be no more than the potential value of its property assets. The development of a new enterprise is covered in Section 3.8.3, and its future value will depend very much on the quality of its operational management, and the skill of its financial management and advisers.

3.7.3 Taking decisions on business options

The style of top-management decision making will be influenced by the size of the enterprise:

- the large organisation is slow to change and may continue too long in one direction, but it can command large resources
- a smaller dynamic company has flexibility and mobility, but is likely to be resource-limited

A large corporation will probably have a headquarters group that can examine business options in depth. The Strategic Planning Society publishes excellent papers and reviews of this type in the journal *Long Range Planning.* A smaller entreprenurial company will not employ planning specialists, and must depend upon the judgment and 'gut feel' of a few of its key people.

The ways of formally assessing the market were outlined in Chapter 1 (1.7), and a simple example of a financial assessment of a proposition was described in Section 3.7.2. At any point in time enthusiasm for any single venture should be moderated, before a final decision is taken, by a review of the alternative options:

- benefits from proceeding with Project X
- commercial consequences of *not* doing so
- identification and assessment of other alternative projects

A feasibility study requires a team effort by the senior people in all functions of the company – marketing, technical, production and financial. Whatever the attractions of a new venture, the finance man is the longstop who must identify the key factors of cost effectiveness.

At the early stages quite approximate computations will highlight the more or less attractive alternative propositions, provided that no significant factors are overlooked.

An effective technique is for the key team to get away from the office, for a full day or weekend, at a country retreat in an informal atmosphere, to work through the decision processes.

A medium-sized firm that makes effective use of its executives as the planning team has an advantage over the large corporation, whose central planners may be too remote from the action in the operating departments.

An agenda for the team session could be as follows (with notes on the method of approach):

A. Orientation of Objectives

A.1 What market sectors have been identified? (accept at this stage all alternative proposals, even unlikely options, and rank them later)

A.2 What position should the company aim to occupy? (e.g. market leader, minority speciality supplier)

B. Make a 'SWOT' analysis

Identify and list out the Company's:

B.1 Strengths
B.2 Weaknesses
B.3 Opportunities
B.4 Threats

} both internal and external

C. *Choose 'top 5' options*
List the possibilities from *A*, in order of perceived feasibility and preference, as judged by the team on the basis of the factors in analysis *B*, choosing a managable number, such as five.

D. *Develop scenarios*
From combined judgment of the team, put together an imaginative scenario of the most probable growth pattern of each option

E. *Evaluate options*
Put cost, sales price and volume estimates against each of the five scenarios (using simplified indicative estimates that are the best judgment of the team).

F. *Profit plans*
Present the five scenarios as investment/profit plans, over a period, such as a duration of three years (preferably including risk factors, which will provide upper and lower limits for the outcome)

G. *Re-rank options in terms of financial results*
Present the five options in recommended order of preference, together with explanatory notes.

At this stage the team will have developed a plan suitable for submission to the Board of Directors, whose reaction could well be to authorise expenditure for initial exploration of the 'top 3'.

The engineering and accounting members of the team together have the competance to make adequate quantification and costing of these options, without complex computation. At this stage they are looking for the orders of magnitude in differences in ranking the options, not precise figures. Any attempt at high accuracy is illusory, because of the uncertainty inherent in the projected scenarios.

The following references on decision making will be helpful: Blanning (1980), Amara and Lipinski (1983), Leemhuis (1985) and Baker and Kropp (1985).

3.7.4 Break-even analysis

Break-even calculations are readily expressed as equations and modelled on a microcomputer, and it is worthwhile to set up a program if repetitive evaluation is needed.

Graphical presentation is informative because it demonstrates visually the sensitivity or elasticity of the optimum value, and how critical it is.

An elementary example is the decision whether to buy, lease or rent a 35 cwt delivery van, price £6000 (the following assumes average mileage, and ignores cost of driver and fuel).

Buying

Initial cost, less residual value after 3 years = £4500		
Written off over 3 years	£1500	annually
Interest at 10% on £6000	600	
Tax and insurance	500	
Tyres, repairs, maintenance	700	
	£3300	

3 year leasing agreement	£2400	annually
Tax and insurance	500	
Tyres, repairs, maintenance	700	
	£3600	
Rental		
Weekly rate (£150)	£7800	
Monthly rate (£500)	£6000	

Fig. 3.7 illustrates that the break-even point compared with buying is 22 weeks for weekly rental and 6·6 months for monthly rental.

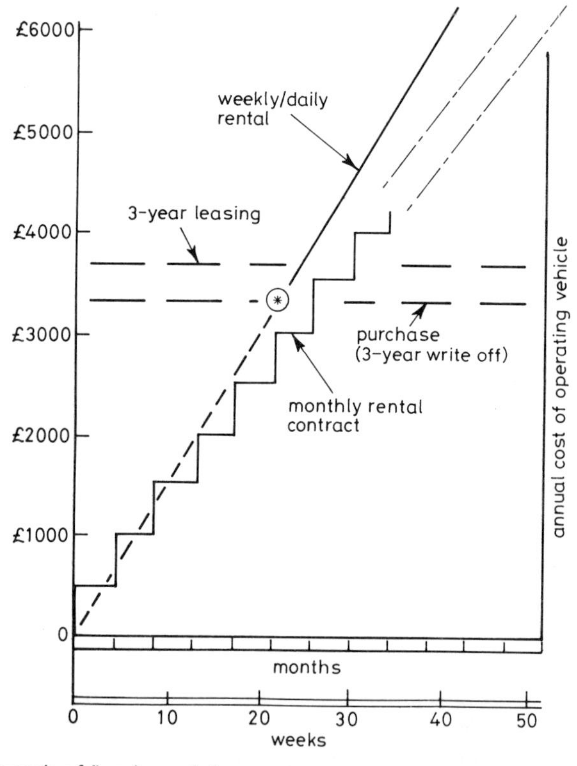

Fig. 3.7 *Example of 'break-even' chart*
Annual cost of operating an additional vehicle (35 cwt van): cheaper to hire as required than purchase ⊛ if below 22 weeks use per year

If there is a varying requirement for one additional vehicle from time to time during the year, occasional rental is attractive. Leasing has no direct advantage, except that it avoids tying up the company's own capital.

Fig. 3.8 is from British Standard BS6143 'Guide to the determination and use of quality related costs', and illustrates how investment in prevention can reduce failure costs, and hence total costs. It is then a matter of commercial judgment to continue beyond the optimum point in order to secure a competitive edge in quality.

The book 'Quality is free' by Crosby (1979) describes the procedures developed at ITT which dramatically reduced scrap, rework and warranty claims at less net cost than the amounts saved. Another example from the experience of Mullard Ltd., quoted by Cohen (1986), was that 30% of operating expense arose from failures of one kind or another.

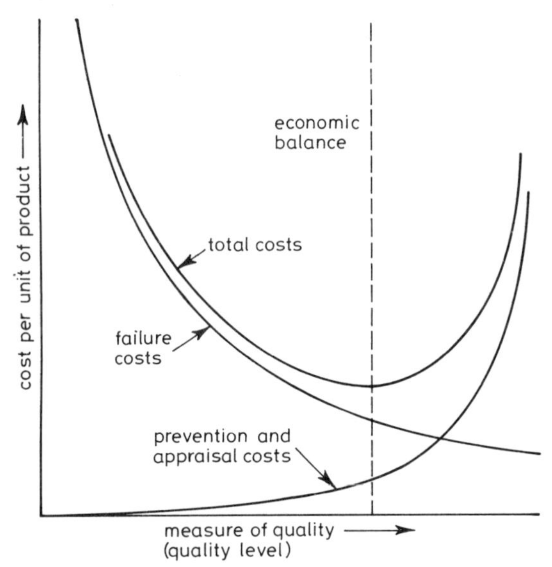

Fig. 3.8 *Quality cost model (from BS6143: 'Guide to the determination and use of quality related costs')*
The prevention costs are offset by the reduction in failure cost, resulting in higher quality at lower total costs (hence Crosby's 'Quality is free' theme)

More complex problems of cost optimisation can be solved by computer simulation, and many examples will be found in the literature of *Operational Research.*

Useful books include Anson (1971), Coyle (1971) and the Penguin books Battersby (1966) and Sizer (1985).

3.7.5 Standard costs and marginal costing

Future business options can usefully be assessed in terms of orders of magnitude, between a wide range of alternatives. This approach is complemented by a much more detailed control of current day-to-day operations.

In the continuous and series manufacture of products, a hierarchy of cost controls provide management with feedback from planning to execution. The control instruments are:

Budgetary control
Comparison of actual performance with budget plan

Standard costing
Target values of detailed costs, consistent with budget
Variance reporting
Completing the feedback control loop to management

These variance statements usually start with analysis of the operating profit variance, and disect the causes stage by stage through all elements of revenue and cost.

The formal definitions are given in 'Terminology of management and financial accounting', published by the ICMA. The techniques have unified into a single system, since the first definitive paper 'The proper distribution of establishment charges' published in 1901 by A.H. Church, whose ideas were quickly adopted both in the USA and UK.

A typical application of budgetary control will include a formal monthly review of the year's manufacturing programme and budget. Cost information will be classified and identified by a coding system, and analysed three ways:

Location
'Cost centre', department, production unit or process
Nature of expense
Wages, materials and certain other expenses
Product type
Type in product range, or a specific order

Fixed expenses are separately budgeted, these being unaffected by the level of activity. They are then allocated as an 'overhead' onto the costs.

The earlier simple concept of 'fixed' and 'variable' costs becomes less realistic as the level of automation and capital investment rises relative to the man-hours input. In a workshop using mainly manual skills rather than machines, the 'overhead' charge might be in the range 100–200% of the direct labour cost. With considerable capital investment in equipment, this ratio becomes 600% or more and loses its true meaning.

'Semi-variable' costs are also identified, being a combination of fixed and variable, or the product of a complex equation. They will include a step function, when an extra machine or additional shift has to be brought into use. This is illustrated in Fig. 3.9.

'Standard costs' are the reference datum, for a product or component made in the programmed quantity, using planned amounts of resources (materials and labour).

'Variance statements' are the error signals to alert those immediately responsible to the need to investigate, and where possible take corrective action.

The system works best for the control of semi-automatic and continuous processes. For example, in the production of wires and cables, and in textile spinning, the process is 'macbine controlled'. It should run close to the 100% capacity of the machine, if the technology is right. Any variance will point to:

- technical problems is setting the machines

- quality problems with raw material
- abnormal levels of scrap

The net cost of scrap will be of special importance where the materials are worth more than the added value of the process.

In construction projects, which are essentially 'one-off', standard costs are not applicable, except where an element of the work is repeated many times, as in

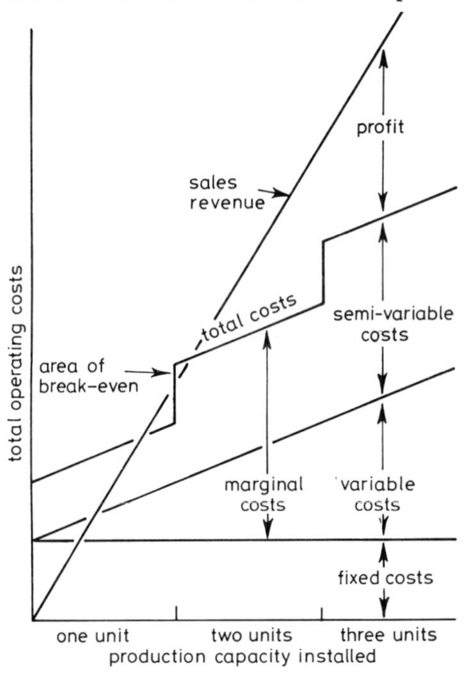

Fig. 3.9 *Illustration of marginal costs*
There are fixed overheads, and semi-variable costs that rise in steps as additional units of capacity are brought into use, either extra machines, or the additional fixed costs of adding a second or third shift: in this example, real profit is achieved only when the second unit (or shift) is put to use

placing piles, or routine installations. Here management should expect to gain from the 'experience curve' effect outlined in the next Section.

Variance reports will usually be issued to the supervisors immediately responsible, daily or per shift. With modern data-processing facilities the reports can be in real time or hourly, and this is worthwhile for controlling a continuous process.

The 'cost centre' accounting concept allows a supervisor to optimise within his own area of authority, in relation to overall cost. Responsibility is taken further when a department is treated as a 'profit centre', and regarded as a self-contained business.

The manager then has targets for surplus or profit, this being the difference between the value of the product and his gross costs. When his product is only one intermediate stage in a chain of production, the 'transfer price' at which his product is valued may be somewhat arbitrary.

'Marginal costing' is used as a guide to pricing, particularly in competitive situations. At the budgeted or normal level of output, each unit sold makes a contribution to total fixed expenses and profit, thus:

$$(\text{selling price}) - \begin{pmatrix}\text{labour + material +}\\ \text{other variable costs}\end{pmatrix} = \begin{pmatrix}\text{fixed}\\ \text{expense}\end{pmatrix} + (\text{profit})$$

Below the break-even point there is no profit, and there is less than full recovery of fixed expenses.

These accountancy conventions are a convenient method of differential analysis, but they can be misleading in isolation, as it may not be appreciated that competitors have a different cost structure, and hence different profit thresholds and marginal costs.

This is one reason why a small company can often undercut the prices of a Division of a large Group, and why a buyout of such a Division by its management team can be successful, when within the Group it was regarded as a loss maker.

Another aspect of 'standard costing' is that by contracting to obtain components or services from suppliers, a company removes uncertainty and guarantees the control of these costs for the period of the agreements. Using internal resources the company would have to carry this risk itself.

When production facilities are owned at 'historic cost' they represent only a modest fixed expense. A prudent company will plan, budget and price in terms of 'replacement cost', and put the surplus earnings into reserve. If this is not done and sales expand, new plant capacity will have to be brought with new money, moving the break-even point for profitability to a considerably higher level. In addition, there is the risk that the market will not absorb all of the increased capacity.

Computer-aided manufacturing facilities are capital intensive, and require additional staff training. Radical changes in practices are desirable to ensure profitability, in particular to adopt two- or three-shift working, and to buy no more than the minimum number of machines.

There are potentially 168 machine-hours/week, and to operate a large investment in machines on only one shift of 35 hours is hopelessly uneconomic. A method used in Japan is to run machines on night shift unmanned, arranged so that they shut down if there is any problem, which is then cleared by the morning shift.

The classic text on standard costing is Batty (1975); it is also covered in most general texts, and by Sizer (1985) in his chapters 3 and 5.

3.7.6 Value analysis and quality systems

The difference between cost and value has long been recognised: a century ago Oscar Wilde wrote that a cynic is one who knows the price of everything and the value of nothing.

Lewis (1969) has recorded how the General Electric Company in the USA assigned Lawrence D. Miles to promote cost reduction by examination of alternatives. By 1947 he had developed a systematic approach based on the identification of unnecessary costs, and named it 'value analysis'.

Miles found that this required much liaison between departments, and so he introduced the team concept to secure more effective co-operation. It was estimated that GE spent nearly $1 m. in developing the techniques, but thereafter saved over $200 m. in the next 20 years.

In 1954 the US Navy Bureau of Ships took up the concept, and called it 'value engineering', with the emphasis on cost avoidance at the design stage. It was generally adopted by the US Department of Defense in 1961.

In his book 'Techniques of Value Analysis and Engineering' Miles (1961) published the definitive material from his training programmes. The Society of American Value Engineers (SAVE) was founded in 1969. General Electric's work became known in the UK following a visit to Schenectady in 1952 by a group from the Anglo-American Council on Productivity.

Several British companies immediately put the ideas into practice, and the Value Engineering Association in Britain was founded in 1966. The early UK enthusiasts included GEC, (Witton), AEI, Philips Electric Industries, Rotax, and then several of the aerospace companies, and the Production Engineering Research Association (PERA).

In additionto Mills's classic book, there are those of Gage (1968), Oughton (1969), and Radke (1972). Many mechanical-value-engineering examples are included in a paper by Conway (1963). Contemporary with Miles's work at General Electric was a parallel programme which led to a book 'Total quality control: engineering and management' by Fiegenbaum (1961).

Thus, the subject area has developed in an inter-disciplinary way, and interest has been shown over the years by the professional institutions and specialist associations:

- IMechE, IEE, ICE, and RAes
- Institute of Production Engineers
- Institute of Purchasing and Supply
- Institute of Cost and Management Accounting
- Value Engineering Association
- Institute of Quality Assurance
- British Productivity Council

The underlying concepts of integrating a company's functions appear in the British Standard for Quality Management Systems, BS 5750:1981: this is also the model for an international standard ISO 9000, which potentially will be adopted by 70 countries. The historical basis is given by Johnston (1988).

The essence of value and quality systems is that they require a combined effort by the main organisational functions of an enterprise, with contributions at each level in each function. The details of technique will have to be interpreted to suit the circumstances of each industry: they can be applied successfully to such diverse fields as mechanical engineering, food industries or software preparation.

Designers and manufacturers have the capability to select an optimum relationship.

performance/cost

and will usually be market-led, to meet the needs and interests of the users. Performance implies not only the technical specification, but also reliability, and esteem values such as appearance and fitness for purpose, which in combination represent quality.

A company will aim a product or service at a market niche for which there is an appropriate price level, so trade off can occur between performance and price. This is most obvious in the field of personal transport, where the price range of cars is at least £5000–50 000.

The Japanese have shown during the last two decades that by wholehearted adoption of quality management techniques (originating in the West) it is possible to offer levels of reliability up to ten times better than has been thought possible in a particular price range. This has been done not only in consumer electronics, but in cars, room air conditioners and domestic appliances.

In project and construction situations, there is an additional trade off option available:

quality/cost/time

A major construction project, taking several years before revenue accrues, will incur considerable interest charges: a valid option is to 'buy time' by using overtime or two-shift work.

In series manufacture of products the opposite effect is seen in the 'experience (or learning) curve'. Fig. 3.10 is an example from the aircraft industry, illustrating that man-hours and cost per unit drop as the cumulative number increases. It was observed that labour costs for the B-29 aircraft dropped 20% for every doubling of output: on log-log axes this relationship is represented by a straight line.

Credit for the original concept goes to the Commanding Officer of McCook airfield, Dayton, Ohio, and first publication was by T.P. Wright of the Buffalo plant of Curtiss-Wright in 1936. It is the residual quantity of 80% that experiences the reduction of one fifth.

The concept is valuable, but there has been confusion in applying it. The important principle is that it is normal to expect continuing improvement, and this is at variance with the classic ideas of work study and measurement, dating back to Taylor in about 1900.

The '80% rule' has often been taken too literally: that value was observed in a particular industry with a high element of operative tactile skills in the assembly stages. Other industries will show their own characteristic percentages. With high degrees of automation, the 'learning' is transferred from the manipulation of the product itself, to operation of the continuous automated system.

The point is made that to use the experience-curve effect for planning and decision making by management, there must be comprehensive records on historic costs, and recognition of points of discontinuity where there were significant changes in the product or the process. The learning effect on people reduces man-hours, but in parallel there will be an accumulation of technical improvements, dependent on the level of input of resources to development and production engineering.

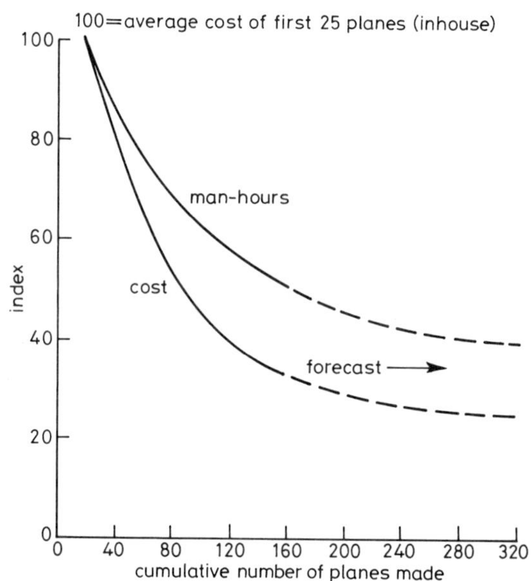

Fig. 3.10 *An example of the 'learning curve effect' from the aircraft industry* (*from F.14 by Grumman*)
After making 80 planes, man-hours were 24% per unit less than after 40 units: again at 160 units it was 24% down on the 80 unit level: a further 24% reduction is forecast for the 320 unit level

For a given manufacturing task where there is an automation/manual option, the long-term trend will be that the automation costs become less per unit produced, and that manual work costs more. In order to plan ahead for more than a year or so, it will be necessary to feed into the cost equation forecast of economic trends.

It is a sounder basis to relate the experience curve to 'added value', rather than total costs, thus separating what is within the internal control of the manufacturing unit from that element of basic cost which is determined by overheads, and by the cost of procured materials and services.

These two factors can then be attacked separately: in an expanding or long-run production of a particular product economies can be sought in the overhead and administrative costs and, in addition, this total is spread over more units. The cost of materials, bought components and services can also be expected to fall below the rate of inflation, if there is active collaboration between the procurement function and the suppliers. Although the selling price of the product may not actually fall, attention to these aspects can go a long way towards holding prices level during periods of inflation.

The BIM Checklist no. 53 on 'Value Analysis' is useful, and it includes a reading list. The basic approach of value engineering is to identify by cost analysis the 20% of elements that contribute 80% of the costs, and then attack these intensively.

The approach in 'reliability engineering' is similar, but directed to the 20% least reliable elements. A 'brain storming' approach is often effective, in which a team representing all functions feeds in their experience and their ideas (however wild). These are subsequently investigated methodically, ranked in value, and priorities agreed for implementation. After an interval, the results and the current situation can be assessed again, and the process repeated in order to achieve another round of improvement.

There is a BIM Checklist no. 77 on 'Managing Purchasing', and there are helpful reviews by Cooke (1967), Conley (1970) and Sallenave (1985) of the application of experience curves.

In the UK there have been experiences of heavy overspending on major projects – for example Concorde escalated from a feasibility study of £0.7 million to a total of £2000 million described by Hall (1980). More recently there has been close involvement of industry in the formulation of requirement specifications for the equipment required by the armed services and in the management of major contracts (MoD, 1983). In USA the 'zero base budgeting' concept was first developed in the electronics industry by Texas Instruments (Pyhir, 1973), and some applications are reviewed by Bragg (1982). On the initiative of President Carter, while Governor of Georgia, the technique was adopted in the public and federal services. The underlying principle is that management will regularly re-evaluate each activity, to see if it is still necessary, and will reallocate priorities in the light of changing needs.

3.8 Financial management and sources of finance

3.8.1 Managing cash flow and profit

The financial chief executive of an enterprise may have a variety of titles such as Financial Director or Chief Accountant. The American term 'Comptroller' (or 'Controller') is the more expressive.

The Financial Controller is actively looking at small differences (or variances) from the target figures, and initiating corrective action. Companies normally aim to achieve a steady record of profit, year by year. A good deal of sub-optimisation and fine tuning is required within the target budget, month by month. The financial executive carries both short-term and long-term responsibilities, including:

- profit planning
- interpreting results of operations, to the several levels of responsible operating management

- controlling the current cash position
- planning the company's financing and investment
- meeting the requirements of corporate taxation, etc

During the initial growth of a new business venture it is very desirable for a profesional accountant to be included in the management team.

A company will normally have several sources for the two-way flow of the funds that provide the working capital:

- bank-balance/overdraft facility
- tax-liability/tax-reserve certificates
- net balance, creditors/debtors
- sale/purchase of assets
- other extraordinary receipts/expenditure
- medium-term loans negotiated or repayable

These elements will be scheduled in detail, as a projected cash flow through the year. Actual cash flows are compared (probably on a daily basis), and the necessary short-term adjustments made.

This will be done by advancing or retarding planned commitments, taking up agreed overdraft facilities, or alternatively putting surplus cash where it will earn interest.

A business will agree with its bank a level of overdraft facility, as a means of smoothing out the inevitable short-term peaks and troughs of income and expenditure. However, the bank manager will be quick to point out that the purpose is temporary accomodation, and should not be the means of financing the company.

Bank loans are a relatively expensive way of securing working capital: if a company is well regarded it can source its working capital in a series of steps, starting with the permanent share capital, supplemented by long-term and short-term loans at favourable interest rates. These will be repaid on due dates from revenue. Thereafter the company may choose to build up deposits and investments to fund future projects. A company that accumulates cash, but does not use it to further its business objectives, becomes a target for a take-over bid by others who believe they could manage better these assets.

A cash budget must be monitored daily in order to set planned expenditure commitments against due dates of payments against invoices. Cash surpluses can be placed on the money market as short-term deposits to earn interest. As this can be done by electronic funds transfer (EFT), overnight deposits can be made when large sums are involved.

An important aspect of cash budgeting is credit control. A company will make clear to its customers its 'Terms of Business', including the period for settlement, which is likely to be in the range 4–13 weeks, and typically 45 days.

New customers are required to establish a credit rating. Regular customers will quickly be reminded if they become overdue, by a systematic process of follow-up.

For this purpose, it is usual to have a credit control section within the Accounts Department, which also authorises the acceptance and delivery of new orders. In a contemporary computerised system this is done with a signal marker against the customer's particulars, when they are called up on the Sales Department's VDUs.

Depending on the nature of the market, a supplier may be more or less tough in the application of his terms of business, but this must not be done by default. Any concession has to be agreed at an appropriate level in the management, in accord with laid down policy and rules. There may be two or three points of contact with customers, the sales people, design engineers and accounts, which are kept separate, but mutually informed. When account problems arise, the sales people and engineers can stand back and take the role of the 'nice guys'.

The process just outlined will also be going on inside the suppliers to the company, who may be more or less generous over the terms of payment. There are some trades where bulk materials are converted, or components assembled, in

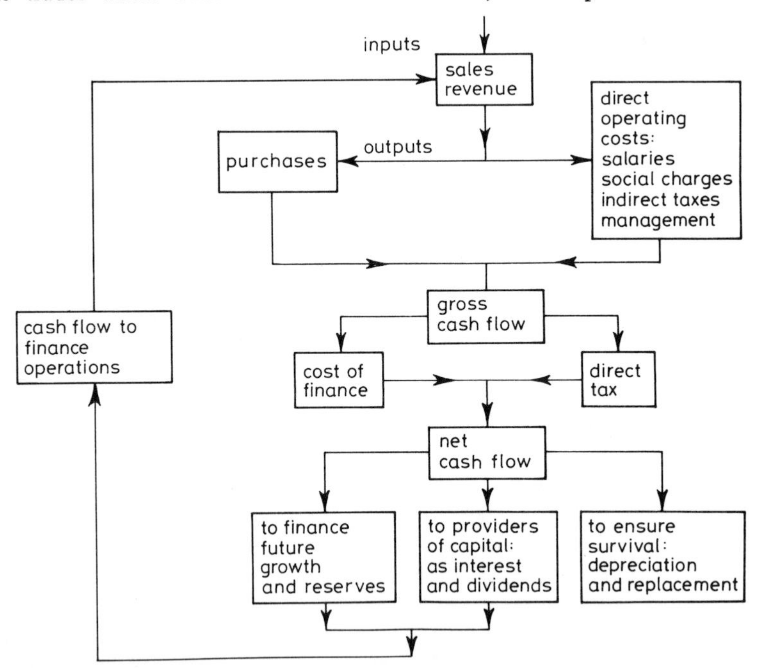

Fig. 3.11 *Illustrating the several streams of cash flow*

which there is a longer period of credit for the materials than is allowed to the end customer. Consequently, the convertor's business is partially financed by his materials supplier.

The Finance Department is responsible for collating and distributing the budgetary control control information down through the organisation, and for reporting the overall results upwards to the Chief Executive, and to the Board of

Directors. This leads on to production of the information in the annual accounts, which will be approved by the company's Auditors.

They, as expert accountants, will act to a greater or less extent as advisors to the company.

The development of the financial reporting and control systems is the responsibility of the Financial Director or his Head of Management Services, and consequently they usually operate the central computer services. As computer networks and intelligent terminals become more general to serve all functions, this historic relationship may change.

Fig. 3.11 illustrates the several streams of cash flow. Useful books on financial management systems include Neuschel (1960) on the basic practices in USA.

3.8.2 Capital project appraisal and control

We have seen that management of cashflow is an essential financial activity and, in any dynamic company, there is a comparable need to plan and manage its 'capital flow'. The requirement may be to finance a steady rate of expansion, or to undertake a major project of re-equipment, setting up a new plant, or acquiring another company.

The degree of need will depend upon the type and size of the business, as in these alternative scenarios:

Small family business
: Working capital provided out of revenue, and ownership in hands of the family: they can take money out as salary, 'perks' or dividend, whichever is the most 'tax efficient' (the phrase for legitimate avoidance)

Small new venture (see Chapter 6)
: With potential for growth in an innovative or high technology field: there is a potential for high profit, and inflow of capital is essential to finance growth

Medium-sized established business (see Section 3.8.3)
: Medium technology and medium growth rate: to maintain its market position it will have to invest in updated manufacturing plant and in product R&D to maintain its position in the market; the source may be revenue, or by attracting additional capital, or by securing medium-term loans

Subsidiary of a large Group
: Dependent on Group HQ for policy decisions and supply of capital: must develop a strong case in order to secure authorisation

Large Group – carries out portfolio analysis
: Group has funds at its disposal for investment, but will explore a wide range of possibilities before reaching a decision on where, what and when

When outside finance is sought, the questions asked of the company can be very searching. The financial community takes a pragmatic and relative view of their opportunities for investment, and can rank the attractiveness of one proposition against others. Walton (1987) has described the roles and attitudes of the City institutions. When a company is proposing to reinvest its own money in a future commitment, it should be equally critical in its appraisal, and minimise its emotional involvement in any 'pet' project or nostalgic dream scheme.

Large corporations finance ventures or projects, usually by proceeding in stages. Initially there is an investment in a feasibility study before making a commitment to proceed. In computer, telecommunications, vehicle and aircraft development, where a great deal is at stake, three alternative approaches may be taken through the stages of R&D, design, mock-up or pilot production, before the final choice is made. These preliminary costs are an insurance against failure, and may amount to only some 10% of the final cost of tooling up for manufacture and making the marketing launch.

Major investments in oil and mineral extraction are approached in the same way, through escalating stages of feasibility studies. A project may be taken through all the stages of planning, major design and invitations to tender. If international political or market conditions change, the whole scheme may then be put on the 'back burner'.

At a colloquium on the systems management of such projects, a speaker from BP International, Francis (1986), said that, in their experience, it was wise to spend more in these earlier stages to avoid unfortunate surprises later, and the 10% figure was mentioned. This can be a considerable outlay as some projects now involve up to £500 m. and, if embarked on, will have a 10-year life.

Large companies cannot afford not to speculate in the investigation of alternative projects, if they are to maintain their position for the future.

3.8.3 Sources of finance and Venture capital

Finance is no problem for an established public limited company (plc) with a track record of profitability; the financial community, anxious to invest soundly, are pleased to oblige. But it less easy for a new venture to get its foot on the financial ladder.

year	activity	working capital	source	turnover	staff
1	launch	£ 25,000	Bank	£ 10,000	½ aver
2	expand sales	£ 50,000	Bank	£ 80,000	2 aver
3	additional products	£100,000	Bank	£160,000	4
4	expand sales	£200,000	Bank + Vent Cap*	£320,000	7
5	new products	£500,000	Vent Cap*	£750,000	12
6	expand sales	£700,000	Vent Cap	£ 1.5 m	22
7	new products	£1.0 m	Vent Cap	£ 2.5 m	35
8	consolidate	£2.0 m	USM	£ 3.5 m	55
12	go public	£6.0 m	Stock Ex	£10.5 m	150

*or Business Expansion Scheme. (all at constant pounds)

Fig. 3.12 *Sources of short- and long-term finance appropriate to the phases of growth of an enterprise*

Fig. 3.12 summarises the sources of short- and long-term finance, both for a start-up company and an established one.

Professor Rhodes (Chapter 6 (6.2)) has outlined the scenario of starting up a small company. He makes the point that technology companies, the ones in which entreprenurial engineers are likely to be involved, are a special case. Less than 1% of all the companies launched in UK are involved in technology.

In the case history he outlines, it was possible to start the company as a part-time activity from his base at the University of Leeds. So, without any large initial investment, it was possible to demonstrate that there was a good profit potential. Secker and Macrosson (1981) have described a number of other ventures that began from a University base.

Rhodes obtained the backing of a venture capital group, and then had the opportunity to expand into the much larger market that was identified in the USA, and was able to look forward to flotation as a public company. There are four other case histories in Section 6.

A study by Rothwell and Zegveld (1980) underlined the national contribution to innovation that is made by small and medium sized manufacturing firms. Since the Bolton Report of 1971 on Small Businesses (Sir John Bolton, formerly of Solartron), there has been increasing Government encouragement for small firms, via favourable tax treatment, assistance schemes funded by the Department of Trade & Industry, plus the availability of venture capitalists.

There are now several directories available, including a Guide to sources of finance for the smaller company from the Institute of Directors (now in its fifth issue), Bienkowski and Allen (1985), one on venture capital for the electronics business by Young (1985), and a guide from the National Westminster Bank. Several others are listed by Windsor (1984), including reference to some 20 official schemes. NEDO (1986) have issued a review of external capital for small firms.

There is a tradition in USA of engineers and academics starting up a business in their garage, and Bill Hewlett and Dave Packard began this way in 1938, assembling an audio-frequency generator based on their invention of an *RC* oscillator circuit. According to *Fortune* magazine the founders have a combined personal wealth of \$3 billion, and their company employs some 70 000 and has a turnover in excess of \$6 billion (1985). H-P has also been given the accolade of excellence by Peters and Waterman (1982).

In other countries the pattern varies. Lyons (1976) describes how Akio Morita built up Sony Corporation, from humble beginnings in 1945. The common factor between Japan and Germany today is that their banks provide a high proportion of the capital requirements, and in return provide expert financial supervision. They expect detailed business projections and feedback of financial results, exercising a tight discipline on performance.

They take a long view, and do not insist on immediate returns. In the USA and UK the influence of the Stock Exchanges puts the main emphasis on the year to year results, which may prejudice longer-term expenditure, as on R&D.

This pattern now applies in the UK to new ventures. The professional sources of finance will require to see a well developed business plan, even for a very modest enterprise. If it is presented spontaneously, they will be favourably impressed, as the banks receive many supplicants who have made their calculations on the back of an envelope.

Banks are more concerned with the quality of the applicant as a manager and a person than with the technology in which he is involved. They are impressed if after a year the results fulfil or exceed the plan, and thereafter their confidence grows with the regularity with which intentions are converted to results.

An enterpreneur, starting from scratch, has to work his way through a hierarchy of sources of capital, and this can take a great deal of his time. When he is also spearheading a high-technology development, he is at risk of being personally overloaded over a long period; so his personal attributes must also include stamina of the long-distance runner.

As an example, suppose someone has identified a specialist market for a digital control device, based on novel concepts, and there is a prototype that can be demonstrated. Fig. 3.13 represents a 'Business Plan' for the first 12 months, and would be backed up by a descriptive prospectus. ICSA (1985) describe how to form a company, and Rich and Gumpert (1985) have written about Business Plans.

For this first year the originator requires enough working capital so that he can arrange for the commencement of manufacture and marketing, during the first six months. Then, in the second half of the year, he would give up his previous job and promote the new venture full time, with some part-time assistance. For the type of product mentioned he will only have to assemble and test, using purchased components, and either subcontracted hardware, or a standard system of PCB card enclosures. He can finance the latter on his credit cards, by buying from a distributor.

By the end of the first year he would hope to have made some sales, have gauged the market, and would aim at moderate expansion in year two: he would also look for some related diversification or application development contracts to carry him into year 3. Tasks for large companies or Government Establishments are useful to stabilise the turnover, and the financial world takes a good view of customers' accounts when they have prestigious names.

Unless private capital is available, he will attempt to cover his first year's needs for working capital from his bank. The manager will talk about colateral for the bank loan, so it will probably be the entrepreneur's house that is at stake.

Discretion of a local bank manager is limited, but he will introduce his client to the specialist departments for small businesses that are usually available at District Offices or Headquarters. Both Barclays and National Westminster have gone a long way with services of this sort. Barclay's approach has been described by Alexander (1985) and by Bullock (1986): the traditional bank overdraft, payable on demand, is now supplemented by medium-term facilities going out to ten years, but at low risk, with security required.

A different and complementary form of high-risk finance is provided by the venture capitalist. The methods of 3i group have been described by Wesley (1985).

	year 0 prior	Jan	Feb	Mar	Apr	May	Jun	Jul	Aug	Sep	Oct	Nov	Dec	year 1 totals	Q.1	Q.2	Q.3	Q.4	year 2 total
		1987 – year 1													1988 – year 2				
Sales of product																			
units A		–	–	–	10	20	20	30	40	40	50	50	50	310	200	240	270	300	1010
B		–	–	–	–	–	–	–	–	–	–	–	–	–	–	10	40	50	100
income A		–	–	–	–	900	1800	3000	4500	6000	6600	7500	7500	37800	26100	26400	29700	30000	112200
received B		–	–	–	–	–	–	–	–	–	–	–	–	–	–	1300	8400	10000	19700
total income		–	–	–	–	900	1800	3000	4500	6000	6600	7500	7500	37800	26100	27700	38100	40000	131900
Expenditure																			
materials A	3500	350	150	200	174	348	453	681	906	958	1132	1132	870		3132	4000	4200	4500	
and devel. B	–	–	–	–	–	–	–	–	–	–	–	–	314		994	3000	3500	3500	
	3500	350	150	200	174	348	453	681	906	958	1132	1132	1184		4126	7000	7700	8000	
rent	–	–	–	–	75	75	75	75	75	75	75	75	75		225	300	300	300	
postage	50	10	10	20	30	40	45	50	50	50	50	50	50		150	200	300	300	
telephone	150	20	20	20	30	30	30	30	30	105	105	105	105		315	400	500	500	
advertising	–	–	–	–	240	260	260	300	300	300	300	300	300		900	3000	4000	3000	
transport	500	100	100	100	200	200	200	200	200	200	200	200	200		600	1000	1500	1500	
samples	–	–	–	–	20	30	40	50	50	50	50	50	50		150	300	3000	2000	
printing	200	–	–	–	50	50	50	50	50	50	50	50	50		150	400	1500	1500	
contingencies	–	–	–	–	75	75	75	75	75	1000	1000	1000	1000		3000	4000	4500	4500	
salaries	–	–	–	–	500	500	500	750	750	1750	2400	2400	2400		7200	7000	11500	11500	
totals	4400	480	280	340	1394	1608	1728	2261	2486	4538	5362	5362	5414	30153	16816	23600	34800	33100	108316
cash flow (for year)	– 4400	– 4880	– 5160	– 5500	– 6894	– 7602	– 7530	– 6791	– 4777	– 3315	– 2077	+ 61	+ 2147		+ 9284	13384	16684	23584	
cash flow (for period)	– 4400	– 480	– 280	– 340	– 1394	– 708	+ 72	+ 739	+ 2014	+ 1462	+ 1238	+ 2138	+ 2086		+ 9284	4100	3300	6900	
gross profit or (loss) for year	(4400) prior												year 1987	£7643				year 1988	£23584
							Gross profit on turnover						20.2%					17.9%	

Fig. 3.13a *'Business plan' for the first year of an enterprise, with outline budget for the second year*

The entrepeneur starts up in January 1987, having previously sunk £4400 in development in his own time on product A. He plans to give his whole time to the new venture from April, when he makes his first sales. His accumulated cash deficit of £7600 is projected to be cleared by November.

In the second quarter of 1988 he plans to launch product B: having demonstrated that he has a valid proposition, he will prepare a more formal business plan and seek the outside capital he needs for substantial expansion of the business

PROFIT AND LOSS ACCOUNT FORECAST	Jan	Feb	Mar
SALES			
Sales			
COST OF SALES			
materials			
labour costs			
change in stocks			
other direct costs			
OVERHEADS			
manufacturing			
selling			
technical			
administration			
OPERATING PROFIT			
finance charges			
PROFIT BEFORE TAX			

OVERHEADS FORECAST	Jan	Feb
MANUFACTURING		
rent, rates		
transport		
maintenance		
plant depreciation		
SELLING		
salaries, NI etc		
discounts etc		
bad debts		
travel		
printing		
advertising		
vehicle depreciation		
ADMINISTRATION		
insurance		
post, phone etc		
audit, other fees		
cleaning		
furniture depreciation		
FINANCE COSTS		
bank interest		
loan interest		
leasing and HP		

CASH FLOW FORECAST	Jan	Feb
CASH RECEIPTS		
cash sales		
debtors		
VAT net rec.		
sale of assets		
other receipts		
CASH PAYMENTS		
materials – cash		
– on credit		
wages		
overheads		
taxation		
VAT net payment		
capital expend.		
loan repayment		
leasing and HP		
bank interest		
BANK BALANCE		
closing balance		
less open bal		

Fig. 3.13b *Formal 'business plan' forecast, developed from the simplified version of Fig. 3.13a, showing the principal headings of analysis in the three main financial statements, on a month-by-month presentation*
The forecast provides the framework for detailed budgets, against which are recorded actual figures: management attention will be given to the variances

They look at the initial business when it has already demonstrated some potential: in order of importance they look at:

- the product and the market
- the people involved
- the budgets for the first three years
- the technologies involved

Their main attention is to the key people in the team, and their awareness of the market.

Unlike the banks, the venture capitalist is not in there to earn interest on his money. What he will take out eventually as his profit is a share of the equity,

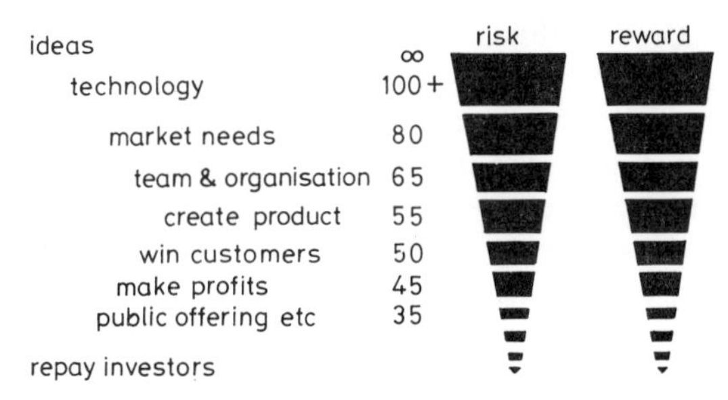

Fig. 3.14 *Life cycle from new technology to profitable product* (*from Taylor, 1986*)
Risk and reward move in step, and the percentages are the annual return on investment that a venture capitalist would expect

quite a substantial share to compensate him for other ventures that fail. Fig. 3.14 from Taylor (1986) of 3i indicates how risk factor changes during product life.

The Unlisted Securities Market (USM) opened in 1981 and now has over 500 companies. It provides a step toward a full listing on the Stock Exchange, but without as great an expense. To enter the USM might cost a company up to £100 000, which is not warranted unless well over £1 m. is being raised. The lower limit for the Stock Exchange would be of the order of £10 m.

To join the USM a company needs the assistance of and sponsorship by an accountancy firm, a broker and a merchant bank. Several of the large accountancies can provide booklets describing more fully what is involved; to name a few:

- Arthur Anderson
- Arthur Young
- Ernst and Whinney
- Price Waterhouse
- Spicer and Pegler
- Touche Ross

3.8.4 Control in decentralised companies and multinationals

Professional engineers or scientists in a smaller organisation have more opportunity to become acquainted with all the functions of business and general management.

They will have informal access to the activities of functions other than their own. In a large Group these other functions become impenetrable.

A multi-company group or a multinational may not consider that the middle ranks need to comprehend the whole picture. Consequently staff are permanently committed to their specialisms, which may inhibit their versatility in the longer term. Integration into large company groups appear to be the trend for the future, and it is well to recognise the difficulties that ensue for individuals.

Particular Groups have developed their own pragmatic organisational solutions, from the very tight control of ITT in the 1960s and 1970s (outlined in Section 3.7.1) to the compact headquarters staffs of GEC and Racal, and the distancing practised by the Hanson Trust, a conglomerate Group where the central role is more that of a banker than direct management.

Executives in a Group subsidiary have to be quite clear about, and accept, the 'rules of the game' in which they are a player, even though these rules at times seem arbitrary. There is nowhere near the freedom of expression and initiative that is available to that manager's counterpart in a comparable independent enterprise. If the opportunity of choice arises, the individual should consciously elect whether to follow a career as a big-company person, or as an entrepreneur.

The more enlightened Groups have made some effort to counteract this polarisation, and the tendency that goes with it to shed some of their more lively people. They arrange for suitable persons to have periods of duty in other functions, and with different subsidiaries, including those in other countries. They arrange some training and indoctrination for those who are in line to be heads of functions, or managing directors of subsidiaries.

It is sometimes said to be a failure by top management if they have to go outside to fill any senior appointment; but probably it is wise to make a minority of appointments from outside, in the interests of cross-fertilisation of ideas between Groups. In the school of 'hard management' this is also done in pursuit of Voltaire's maxim '. . . it is thought well to kill an Admiral from time to time to encourage the others'. He was saying this of England, in the 18th century!

The incoming person is also at risk, for he has to accept the current cultures and disciplines (unless he goes in to head the Group). A newcomer to a Group faces the discipline of a budget monitored by headquarters, and the responsibility for achieving it, or calling for timely help. This back-up can be provided in a Group, but is less readily found by the MD of an independent company.

Multinational companies have to contend with the added dimensions of local political and cultural differences. Both tightly and loosely-knit examples are found, but the general tendency is towards imposing global standards:

technically: for products and processes
politically: to set an operating objective of working within each country's cultural styles, to conform to local practices, and to offer conditions equal to or rather better than the norm.

As mentioned in Section 3.3, harmonisation of financial accounts has been effected within the EEC, and there is a move towards global standards, so that the accounts of multinationals can be presented coherently.

This is of interest, both to investors and for reasons of taxation. A corporation operating in several countries may find tax treatment in one more favourable than in another. Some countries will offer special development incentives, and 'tax holidays' for new ventures.

When products or components are made in one country, but assembled and marketed in another, the issue of 'transfer pricing' arises. This is because a multinational has the opportunity to take its intermediate profits in whichever country it is the most 'tax efficient' to do so. They are likely to be required to justify their internal transfer pricing against prices in the open market, and they may be accused of 'dumping' surplus production at unrealistically low prices, constituting unfair competition (Sizer, 1985, chapter 11).

There is a continuing debate on the effect on the national economy of sourcing products abroad, e.g. consumer electronics manufactured in the Far East. There are two opposing points of view:

- sourcing abroad cuts manufacturing jobs at home
- conversely, a large number of consumers benefit from the lower prices, and competition in specification and quality
- the trading company retains its marketing and distribution business, that otherwise would be lost to foreign competitors

Companies trading internationally have to protect themselves against exchange-rate exposure, which can be a serious risk in long-term commitments. Babbel (1982) has described the implications for engineering companies, and there are facilities through the money markets to 'hedge' the risks.

For a background understanding of the multinationals, it is useful to read how General Motors set up self-accounting divisions and subsidiaries in the 1920s, in the book by Sloan (1963). Also the chequered history of ITT in the versions by Sampson (1973) and of IBM by Rodgers (1986).

In his book 'Small is beautiful', Schumacher (1973) commented that Sloan's great achievement was to structure the gigantic General Motors in such a manner that it became, in fact, a federation of fairly reasonably sized firms, and this formula has become the norm.

3.9 Reading list

BATTY, J. (1976): 'Accounting for research and development' (London, Business Books)

CROSBY, P. B. (1979): 'Quality is free' (New York, McGraw Hill) 309 pp.

FANNING, D. (1983): 'Handbook of management accounting' (Aldershot, Gower) 496 pp.

HARGEAVES, R. L., and SMITH, R. H. (1984): 'Managing your company's finance' (London, ICFC)

ILO (1985): 'How to read a balance sheet' (a programmed book) (Geneva/London, ILO, 2nd edn.)

MOTT, G. (1980): 'Accounting for non-accountants' (London, Heinemann/Pan)
OLDCORN, R. (1980): 'Understanding company accounts' (London, Heinemann/Pan)
POCOCK, M. A., and TAYLOR, A. H. (1981): 'Handbook of financial planning and control' (Aldershot, Gower) 446 pp.
SIZER, J. (1985): 'An insight into management accounting' (Harmondsworth, Penguin, revised 2nd edn.) 525 pp.
VIDEO ARTS: series of video cassetes with notes: 'Balance sheet' 'Working capital', 'Cost, profit and breakeven', 'Budgeting', 'Depreciation and inflation'

Chapter 4

Legal and social obligations

4.1 Summary

Introducing the then far-reaching proposals for social insurance in the early 1940's Lord Beveridge said:

> The object of government in peace and in war is not the glory of rulers or of races, but the happiness of the common man

The same note was struck in the American Declaration of Independence of 1776:

> ...That all men are created equal; that they are endowed by their Creator with certain unalienable rights; that among these are life, liberty and the pursuit of happiness. That, to secure these rights, governments are instituted among men, deriving their just powers from the consent of the governed;...

By the formation of the Constitution in 1787, a framework was established of legislative, executive and judicial powers, which provides the point of reference for any new legal provisions that become necessary.

In Britain, the principle of 'the rule of law' ensures that the Constitution is adaptive rather than prescriptive. When it becomes evident that 'the law is an ass', and 'nonsenses' are generally agreed to exist, initiatives are taken to change the law, but with the inevitable time lag inherent in a democratic consultative process.

In its detailed interpretation, English law is built up from the decisions on test cases, and Scottish law is constitutionally free to diverge. Test cases arise from circumstances not foreseen by the original draughtsmen of current legislation and a number of the notable ones have involved decisions on the consequences of new technology, and the regulation of business behaviour. Particular situations have been resolved through arbitration.

An objective of the European Economic Community is to harmonise the legislation of the Individual States, but national characteristics tend to persist in the general practices. This was summed up by the former Chairman of the US Federal Communications Commission, Newton Minow (1985), who said:

> In Germany, under the law everything is prohibited except that which is permitted. In France, under the law everything is permitted except that which is prohibited. In the Soviet Union, everything is prohibited, including that which is permitted. And in Italy under the law everything is permitted, especially that which is prohibited.

The pattern of history has determined that the British most nearly resemble the French, but in practice with greater discretionary powers in the hands of the judiciary.

Managers involved with technology need to be sensitive to the legal, social and cultural implications of their work. Their understanding should go beyond the 'letter of the law', to the underlying objectives, as in the words of Lord Justice Hewart (1924):

> Justice should not only be done, but manifestly and undoubtedly be seen to be done

The vocabulary of the law is usually less transparent than this statement, and a general reference book such as 'Everyman's' (1981) and Smith (1979) can be helpful in mastering the language.

There is a tradition from the 19th century that a few enlightened industrialists have exercised moral leadership and assumed a degree of social responsibility well in advance of their legal obligations. Their earlier models have through subsequent legislation become the norm for business as a whole.

4.2 Legal responsibilities of a company

An individual who becomes a Director of a company has a role in law quite distinct from the operational function as a senior executive manager, and shares this collective responsibility with all other executive and non-executive Directors who may be members of the Board.

The Institute of Directors (1985) advises that we:

> . . . regard the company, like any other form of business organisation, as a system which links customers and all the parties who contribute to the customers' satisfaction – but a system in which the relationship between the business and those parties is sophisticated and sometimes elaborately defined

Professional scientists or engineers may have little exposure to matters of company law within large commercial organisations, where these matters are in the hands of the Company Secretary, and possibly a legal department. In contacts with outside clients or suppliers, some feeling for legal matters is necessary, and when an individual contemplates joining another company in the capacity of an executive and/or a director, or participating in setting up a company.

There are a very large number of companies in the UK, nearly one million: some are dormant and others are small family businesses, or part-time interests.

Less than 7000 are public limited companies. The public companies offer part of their capital to the public, via the Stock Exchange. Most aspects of company law apply without distinction as to size, although public companies and their subsidiaries represent some two-thirds of the gross national product.

The concept of 'stakeholders' has gained currency in the wider sense to include the impact of the company upon the public domain. Thus 'stakes' are held by the following parties who in some way are associated with the company's activities:

- customers
- suppliers
- employees
- shareholders
- other providers of capital
- the local community
- the national interest

The Board of Directors has collective responsibility in the legal sense, for the conduct of the company, so it rests with the Chairman of the Board to achieve a concensus. The shareholders have the opportunity at the Annual General Meeting (or at an Extraordinary Meeting) to confirm or reserve their support: this is a 'safety valve' that will not normally come into operation.

Within Europe there is commonly a two-tier Board: an executive level and a supervisory level. The banks are represented because they provide a greater proportion of the company's capital, and there has been pressure for a significant number of worker representatives. This approach has been the subject of considerable debate, but, as far as UK is concerned, the main outcome is a move towards including non-executive Directors on the Board, to provide a balancing role.

The Board should determine long-term objectives and receive adequate reports on current operations, not just the formal accounts. It has a responsibility to strike a balance between the interests of the several stakeholders. Directors may be held personally liable and can be disqualified from holding any directorship if misfeance occurs. Under the Insolvency Act of 1985 they can be liable for negligence if they allow a company to continue trading and to run up debts, when its financial position is unsound. There is clearly serious backlash between the deterioration of the financial position of a company, and the appearance of its statutory accounts: Directors can be prosecuted if Annual Returns become overdue.

It normally falls to the professional Company Secretary to advise the Board on all the legal requirements, and the consolidated 1985 Companies Act has over 700 sections. There are many reference books on company law: for a quick overview, and general reference, Morse (1983) and Boyle (1983) are of manageable size. The main aspects covered include:

- origin, nature and development of public companies
- Articles of Association
- Company contracts

- promotion and underwriting of companies
- Company Prospectus: and liabilities for misrepresentation
- dealing in Securities
- Annual Returns, Directors' Reports and shareholders' Meetings
- statutory requirements and duties of directors
- role of the Auditors
- liquidation, voluntary or compulsory winding-up
- overseas companies

4.3 Consumer Legislation and social responsibility

In a book by Harvey (1982), it is explained how the early consumer protection mechanism of the 19th century was linked to local government of the county or borough. This is still perpetuated in the local office of the 'Weights and Measures', now known under the wider title of 'Consumer Protection'. Historically the emphasis was on the correctness of quantities in the market place, it being accepted that a bargain would be struck between the two parties in a sale, with respect to quality and price: hence the warning *caveat emptor.*

With the advent of manufactured and packaged goods, the lay consumer had not the expertise to judge value at the point of sale. In the first half of the 20th century, responsible manufacturers introduced branded goods to assure confidence in quality. Technical products were covered by standard specifications which evolved from often protracted work by sub-committees representative of the producers and users.

The Consumer Union in the United States (founded 1936), and the Consumer Association in UK (1956) were formed to provide independent technical information and comparative testing, and exercised a degree of political pressure. They have moved on deeper into technical assessment of, for example, performance of the telephone service, reliability of TV sets, the life of batteries and lamps, and vehicle testing; and also the performance of DIY and construction products, so overlapping into the industrial field.

A more abrasive note was struck by the movement in the USA of the lawyer, Ralph Nader, who, by using legal skills and a team of aides, has brought considerable leverage to bear on large and small companies.

To counterbalance the greater activity of pressure groups, Baker (1985) argues that engineers must see themselves in the broader context of society as a whole, in order to overcome a general distrust by the public in science and engineering, and to win back their support and interest.

The overall effect of active 'consumerism' and legislation is to increase the supplier's 'compliance costs'. But, as Crosby (1979) of ITT has said in his book 'Quality is free', there is an offset cost reduction in rectification and 'whole life' maintenance costs. The recent attention to 'excellence' in business focusses attention on the benefit to a company of a policy of exceeding their legal obligations for levels of quality and service, and in developing a public image of social responsibility.

A good introduction to consumer legislation is provided by Janner (1979), one of a series of books on aspects of law by this author. The present body of law that may affect the activities of engineers, manufacturers and distributors, in relation to products and services, is built upon the following legislation:

- Sale of Goods Act, 1893
- Consumer Protection Act, 1961
- Health & Safety at Work Act, 1974
- Consumer Safety Act, 1978
- EEC Directive on liability for defective products

The pattern of the practice of law in the United Kingdom is that a need arises for a new piece of legislation, and the appropriate Department of State collects evidence, and the legal draughtsmen put forward their best efforts, which may be modified during passage through Parliament.

4.4 Health and safety legislation

The Consumer Protection Act of 1961 provided the vehicle to enable the Secretary of State to impose requirements for safety, and for the provision of warnings and instructions:

> . . . as to the composition or contents, design, construction, finish or packing of . . . goods . . . or any component part thereof . . . to prevent or reduce risk of death or personal injury

The Consumer Safety Act of 1978 replaced that of 1961, and enables the Secretary of State to:

> . . . make regulations as he considers appropriate for the purpose of securing that goods are safe or that appropriate information is provided . . . hereafter referred to as 'safety regulations'

This is administratively more flexible, as the details do not have to be put through the lengthy Parliamentary process. Safety regulations may require conformance to a particular standard or form of approval, and may determine the standards to be applied in carrying out test and inspection. The supply of goods may be prohibited if they are considered not safe. The IEE (1984) issued a Professional Brief on 'Product Liability', and an informative annual Report is published by the Health & Safety Commission (1985).

A person who contravenes a prohibition notice is liable to imprisonment and a fine, but it is a defence to prove that all diligence was exercised. The full text of the Act is quite clear that the person rather than the company will be prosecuted: this will be the person functionally or professionally responsible, and/or the

Directors. This is the same concept as the personal responsibility of 'The Engineer' in engineering contracts.

Before the Secretary of State proposes to make safety regulations, he has a duty to consult such organisations as appear to be representative of the interests affected. This is an instance where Trade or Technical Associations can be very effective, as a collective voice carries more weight than the pleas of individual interested parties.

The broad sweep of the 1978 Act takes in all forms of trade and business, and applies at every stage in the chain, from design, through manufacture to distribution, retail sale, installation or maintenance.

It is a feature of British law that aspects not foreshadowed by the draughtsman are determined by the rulings in test cases, and hence known as 'case law'. In this area lawyers are turning to data-base technology for more effective information retrieval. Key points of law are known by the names of the litigants and the date; for example:

Harbutt's Plasticine v *Wayne Tank Pump Co (1970)*
Fire broke out in a factory the first time that a new plant was put to use

Vacwell Engineering v *BDH Chemicals (1970)*
Solvents used in connection with semiconductor production carried no warning of the industrial hazards that might arise

Haseldine v *C A Daw and Sons (1941)*
A person was injured in a building, but responsibility was assessed not to be that of the owners, but of the maintenance engineers who had a contract to look after the lift involved

The finer points of such cases are argued by the lawyers, and an initial decision may be reversed on appeal; thus a new point of law is established. In the UK damages are assessed by the Judge, based on the loss that is a foreseeable consequence of the fault. In the USA the decision may rest with a Jury, who may allow a whole chain of consequential damages, so that very large sums could be involved, and the cost of liability insurance is correspondingly high.

A plaintiff faces the risk that, if the claim is rejected, costs can be a severe penalty. The overall risks of pursuing a case must be weighed before proceeding, as the demands on executive time will also be high, particularly, if another country is involved. Even a successful judgment may be 'barren', if the defendant is a 'man of straw' with no resources, or a company in liquidation.

A Professional Brief on 'Health and Safety Legislation' has been prepared by the IEE (1984), and contains much information. It is pointed out that an engineer's liabilities fall under three main headings:

- those arising from breach of the expressed or implied terms of contract
- those due to his negligence or breach of statutory duty
- those which attach to him in his capacity as an employer or as an employee

and the distinction is explained between civil and criminal liability.

Company insurance is dealt with very fully in a Guide issued by the Institute of Directors (1986). The BIM issue a checklist and reading list (No 39) on 'Industrial health care'.

4.5 Employment law

Health and safety at work is both a moral and legal obligation, of concern to every employer and all managers, supervisors and employee. There have been a long series of statutes, some based on 'welfare' considerations, and others directly related to potential industrial problems; e.g. the Alkali and Works Regulation Act (1906), Celluloid and Cinematograph Film (1922) and the Control of Pollution (1974).

Here again, much stems from case law. In one instance an employee was injured using an unsafe method of testing in a substation – the employer was liable because the employee was not aware of the laid down safety precautions (Babcock *v* Brighton Corporation, 1949). In another case goggles were provided for a grinding operator, but no one told him where to find them (Finch *v* Telegraph Construction and Maintenance Co Ltd, 1949).

Management has a responsibility to provide safe systems of work, and have committed a criminal offence if this is not so. First-line supervision have to ensure that each employee is aware of procedures and risks, and is rehearsed in the precautions and the proper reactions in an emergency. Allowance must be made for any problems the employee may have with comprehension, or reading and understanding instructions. Middle and senior management have overall responsibility, and must exercise it by initiating the systems, checking their effectiveness and monitoring the action of supervisors.

Establishments are required to keep an Accidents Book, and employees have to formally report any accident to themselves, if they intend to claim industrial injuries benefit.

The concept underlying recent health and safety policies is that a contract of employment is a mutual agreement, that should not be breached by either side. It applies to all concerned, both management and their representatives, the employees, their shop stewards and Trade Union respresentatives.

A company should provide a rule book as a source of information and guidance, refer to it in the contract of employment, and keep it up to date. It is desirable that copies should be provided to all employees: these rules should be the basis for initial briefing of new entrants. There is a BIM Checklist on 'Formal induction programmes' (No 7).

A good summary of current employment practices and procedures is provided by Croner's Handbook and by Armstrong (1984, chapter 13). The correct disciplinary procedure should be followed, escalating through the three stages:

- oral warning, given informally
- formal warning, backed in serious cases by a written statement of the offence, and consequences if repeated

- final written warning, including advice that any repetition will be penalised

As there is a close connection between the incidence of accidents and carelessness or irresponsible behaviour, strict application of the discipline procedure is important. Most accidents are due to an aggregation of minor infractions of good practice, and are more likely to occur where there is a slack and easy-going atmosphere.

There is a useful book on the law of health and safety by Selwyn (1982). The International Labour Organisation (ILO), based in Geneva, was founded in 1919, and consists of representatives of national governments, employers' and workers' organisations. It develops 'Recommendations' which are submitted to national governments for consideration, and may be finally ratified, or accepted as a guide. Thus, its publications are informative on good practices, and foreshadow future legislation, e.g. a recent review of Workers' Participation (Monat, 1986).

Where persons are employed in attendance on complex plant, such as in process-control rooms, or automated industrial systems, special training is advisable, based on actual simulation of all feasible eventualities. These sessions should be repeated fairly frequently, to condition the responses of those who normally are carrying out monotonous routine operations.

The designers of complex systems should include simulation facilities in the original specifications, as the cost of adding the features at that stage is minimal. Himmelfarb (1985) has prepared a guide to product failures and accidents. The major catastrophies and subsequent Enquiry Reports attract much attention. Lack of knowledge and lack of preparation are generally identified as important contributions to the chain of events. Every business organisation should have a 'disaster plan' with all conceivable eventualities thought through, provisions made, and a clear understanding of who will be responsible for what.

4.6 Contract law

It was Samuel Goldwyn who said 'a verbal contract is not worth the paper it is written on'. The point is that there must be very clear understanding by both parties to the contract, and every effort made to serve their mutual benefit, before the attempt is made to formalise it. Legal phraseology may be unattractive, but is designed to remove any remaining ambiguities. There are two phases in the process of legal contracts:

Negotiation: for the mutual benefit of the parties
Arbitration: if something subsequently goes wrong

Arbitration is available as a mechanism for resolving all kinds of disputes, whether or not a contract is involved, and it is generally more satisfactory than taking the dispute to the courts. It is almost certainly quicker and cheaper, as legal expenses can escalate, and are particularly high where technical issues are at stage, and 'expert witnesses' become involved.

Negotiation is something of an art, and the skills and strategies can be learnt, a useful guide being by Nierenberg (1973). In the technical-commercial field the end point is usually an optimum and cost-effective solution that neither party fully visualised when they started. In one instance in Central America bids were invited specifying a telecommunications cable network, but the winner was the company that negotiated an alternative specification involving microwave relay links along the line of mountain tops.

In another example, civil engineers were invited to bid for a water distribution system over about 100 km from mountain lakes to a city, and successfully offered the alternative of wells bored down to the water table directly below where the supply was required.

Negotiation is best based on a statement of functional needs, leaving it to the bidder to come up with the technical solution: this applies particularly to data-processing applications beyond the range of standard software.

Within a particular field such as engineering construction, Standard Forms of Contract are very familiar, and greatly simplify negotiation, not only because the format is mutually acceptable, but also most of the snags have been ironed out already through 'case law'.

In large projects where many contractors are involved, there is a need for a systematic overall 'contract strategy', of which examples are given by Burbridge (1986). Some client organisations are large enough to 'engineer' and co-ordinate the contractors on their own projects (e.g., BP, LRT, RTZ and CEGB), but in many instances this function is provided by consulting engineers (John, 1984).

When contractual difficulties do arise, particularly through causes not forseen, they can often be resolved between the parties if detected and addressed at an early stage. If it comes to arbitration, there has to be mutual agreement to take this course and accept the Arbitrator's decision. In the context of labour relations, the role of ACAS is well known.

In disputes involving engineering matters, the senior Engineering Institutions offer services, through their Presidents, for the nomination of arbitrators, expert witnesses or consultants, or inquiries can be made directly to the Institute of Arbitrators. The role of the professional arbitrator in the settlement of disputes in the engineering field, has been outlined by Beresford Hartwell (1987).

Morse (1983) reviews Contract Law in his section 6, and Wheeldon (1988) outlines the 'Engineer's' responsibility.

4.7 International requirements and standards

There are two current trends:

- statutory requirements now increasingly refer to technical standards, giving them the force of law
- the EEC countries are moving quite rapidly towards harmonisation of their standards, with fully international standards under active discussion

Under the Strasbourg Convention of 1977, each contracting State of the Council of Europe shall make its national law conform with the provisions of the Convention. Some local differences of interpretation and of case law are likely to remain, together with variations in the specific regulations for safety.

The National Institute for Economic & Social Research sponsored a study by Neale (1970) of the Antitrust Laws in the USA, which is helpful in understanding the basis of competition within that country. In recent years United States policy has been to de-regulate, so intensifying competition in business areas such as freighting, airlines and telecommunications. In the UK, OFTEL is a model of the new approach, and has been described by Carsberg (1986).

Years ago, a famous series of cases under the US Antitrust Laws concerned the right of a patentee to apply retail price maintenance, involving branded pharmaceuticals, General Electric's lamps (1926), and other goods. The original regulation against monopoly action by the big corporations was the Sherman Anti-Trust Act of 1890, in support of the ideal of unfettered economic opportunity. This was described by President Herbert Hoover (an engineer by background) as 'the American system of rugged individualism'. The last major action against a large Corporation involved IBM in the late 1970s, but was eventually withdrawn.

McGovern (1982) examines the rules of international trade regulation, which is a specialised area of law, but one where engineering managers may find a need to have a feeling for the situation. His survey includes customs and trade regulation, tariffs and other restrictions, together with the trade arrangements appropriate to the interests of the developing countries. The White Paper (1982) on standards and international competitiveness set out the UK Government's intention to make greater use of standards as a basis for legislation and public purchasing, and the EEC is preparing Directives for Community Standards that will leave to the individual nations the initiative on detailed technical standards.

The concept of 'intellectual property' provides an integrated approach to what had become a fragmented area of legal protection. The term covers 'written works', patents, designs and copyright. Flint (1985) reviews how the requirements for protection have been broadened by the advent of new media for conveying ideas, such as video recording, computer software, teletext and electronic publishing. Collins (1984) describes the law of the EEC.

The first Copyright Act was passed in 1709, and protected printed works for 21 years, and the most recent legislation has broadened the definition of 'original work' to embrace artistic works, music, sound recordings and broadcast material.

'Electronic publishing' (Flint's sections 30 and 31) represents alternatives in media rather than content, such as alternative methods of transmission or retention (magnetic tape, or IC storage such as read-only memories). An interesting area has been that of video games, recorded on ROMs and incorporating both sound, visual effects and text.

There is a Professional Brief on 'Photocopying' (IEE, 1984), and useful general texts are Russell-Clark (1974) on copyright in industrial design, and Melville (1972) on intellectual property and international licensing. Hearn (1986) writes on industrial licensing, and Reid (1984) on patent law.

Case law includes a number of examples involving large companies, notably the recent Polaroid *v* Eastman Kodak one on instant colour photography, where Kodak was obliged to withdraw from this market. More usually a settlement is made by licensing, rather than face the cost of protracted legal action. In the case of Technograph Printed Circuits *v* Mills and Rockley in 1969–71, the case was taken up to the House of Lords on appeal, and won by the plaintiff. There were notable cases in the 1930s involving thermionic valves: Mullard *v* Philco (screened pentode) and EMI *v* Lissen (variable mu valves). BTH fought a case against Metropolitan Vickers on a method of starting synchronous machines, and recently Racal-Milgo has defended cases relating to modems, brought by Western Electric and by Codex Corporation. The complexity of technical cases can present the legal system with problems, and the judge may appoint a personal scientific adviser.

4.8 Reading list

BERESFORD-HARTWELL, G. M. (1987): 'Arbitration in mechanical and electrical engineering', *IEE Proc.*, Pt. A *134,* pp 343–350.

BOYLE, A. J., and BIRDS, J. (1983): 'Company law' (Bristol, Jordan) 824 pp.

BROWN, J. A. C. (1980): 'The social psychology of industry: human relations in the factory', (Harmondsworth, Penguin) 336 pp.

ENDERSBY, J. C. (editor): (1988), Special issue on 'The Engineer in Society', IEE Proc. *135* Pt.A pp 245–323.

FIELD, D. (1980): 'Inside employment law – a guide for managers', (London, Heinemann/Pan)

HARVEY, B. W. (1982): 'The law of consumer protection and fair trading', (London, Butterworth, 2nd edn.) 424 pp.

HEARN, P. (1986): 'The business of industrial licensing', (Aldershot, Gower) 250 pp.

IEE Professional Briefs on: 'Product liability', 'Health and safety legislation' and 'Copyright: photocopying IEE publications'

MONAT, J., and SARFATI, H. (1986): 'Workers' participation – a voice in decisions 1981–1985', (Geneva/London, ILO)

REID, B. C. (1984): 'A practical guide to patent law', (Oxford, ESC Publishing) 438 pp.

SELWYN, N. (1982): 'Law of health and safety at work', (London, Butterworth)

Chapter 5

Marketing products and services

5.1 Summary

5.1.1 Trends

A comprehensive definition of what is understood to comprise 'marketing' is:

> The management of the relationships between an organisation and its customers and potential customers. Marketing aspects of a corporate strategy may include the mix and volume of products, pricing, distribution methods, guarantees and servicing, advertising and promotion, and sales force management. Market research seeks information on the characteristics of customers, potential customers, and competitors.
>
> (R I Tricker, Director, Oxford Centre for Management Studies)

The difference between 'marketing' as a strategic process, and the narrower function of 'selling' was emphasised by Theodore Levitt (1960) in a paper 'Marketing Myopia'. He summed up his thesis in these words:

> an organisation must learn to think of itself, not as producing goods and services but as buying customers, as doing the things that will make people want to do business with it

This has been summed up succinctly as: 'we are not selling quarter inch twist drills, but quarter inch holes', and this carried with it the implications of performance, quality, customer satisfaction, and durability:

- it is not the price of the drill that matters, but the cost per hole (customer's viewpoint)
- It is not the cost of making the drill, but its value (and price) to the customer (drill maker's viewpoint)

Levitt's 1960 article in *Harvard Business Review* attracted a good deal of attention, and was a turning point in the general conception of 'marketing'. Another article by Adler (1967) on 'Systems approach to marketing' provided a comprehensive background. It was inspired by the 'systems approach', used in the work of The

Rand Corporation for the US Air Force, and their view of the procurement process: marketing in reverse.

Adler quoted as a pioneer in applying the concept the Carborundum Company which, in the mid 1950s, perceived that what customers *needed* was a complete system, using abrasives to shape, cut and finish at minimum cost.

They set about making customers *want* this, by specialising in their technology, so that they were able to take over process operations previously carried out by cutting tools. Their salesmen acquired the status of technical advisers to their clientele, practical engineers who set up and demonstrate the processes, and were welcomed on their visits.

Adler goes on to remind us of the role that merchants have fulfilled for centuries, by quoting from Reginald (*circa* 1170):

> He laboured not only as a merchant but also as a shipman . . . to Denmark, Flanders and Scotland; in which lands he found certain rare, and therefore more precious, wares, which he carried to other parts wherein he knew them to be least familiar, and coveted by the inhabitants beyond the price of gold itself, wherefore he exchanged these wares for others coveted by men of other lands . . .

Drucker (1973) puts 'marketing' into perspective, as one of these seven key functions of business:

- scientific management – key to productivity
- decentralisation of management
- personnel management – fitting people to the organisation
- management development – to meet future needs
- management accounting – information for decision making
- marketing – identifying and satisfying customers' needs
- long-range planning

Many books on marketing techniques have been published, some implying that marketing is the only key function: conversely, most engineers would look for manufacturing technology, development and research in Drucker's key list.

5.1.2 Engineering marketing

Although marketing techniques tend to be associated with consumer goods, services and general advertising, the principles apply equally well to the marketing of engineering products, and to technical and professional services.

In fact, there has been a general convergence of ideas on performance and quality. High levels of reliability are now expected in technical goods, whether for use by consumers, in commercial applications, or by the military. There is a general trend towards adoption of international designs, and international harmonisation of performance and requirement standards. Most businesses, in determining their strategies, have to think internationally, and think ahead.

Most professional people, in the course of their careers, will be presented with options such as a move from research to development, or from design to application engineering, and on to technical marketing or contract negotiation. There will be opportunities for limited or extensive contact with other countries, as a short- or long-term visitor, or to settle.

Other countries, notably Germany, Japan and Scandinavia, have had success in world trade by putting top calibre people into marketing. Unfortunately, for a century or more, selling was perceived as an inferior occupation in Britain, despite the earlier achievements of the Elizabethan merchant adventurers and, later, the East India Company, the Hudson Bay and other great trading companies.

Adam Smith (who died in 1790) wrote in 'The Wealth of Nations' that:

> To found a great empire for the sole purpose of raising up a people of customers . . . is . . . a project extremely fit for a nation that is governed by shopkeepers

Napolean Boneparte, while confined to St Helena, threw back this accusation in his own words: 'L'Angletere est une nation de boutiquiers'.

But our world markets had already begun to slip away well before the end of the 19th century, as foreseen by Disraeli in 1838 when he warned that: 'the Continent will not suffer England to be the workshop of the world' and a century later it was the Far Eastern countries that took up the initiative.

What we still have to counter are the attitudes revealed under the heading 'Advice to Tradespeople' in a book written by Αγωγος in 1834, called 'Hints on Etiquette and the usages of society':

> Society is divided into various orders, each class having its own views, . . . and the literary acquirement of a man of business is necessarily confined to reading the newspaper, . . . and however skilful in his trade, cannot form an idea of that man's mind who has devoted all his energies to science or literature . . . The English are the most aristocratic democrats in the world; always endeavouring to squeeze through the portals of rank and fashion, and then slamming the door in the face of any unfortunate devil who may happen to be behind them

When engineers move into the marketing side of their industry, they are sometimes regarded as having left the profession: it is necessary that the Engineering Institutions recognise that to put expertise to work internationally is a contribution to the community equal to that of their learned-society activities.

5.1.3 Professionalism in marketing management

The function is well established professionally in UK, with its own Institute of Marketing founded in 1911, and over 20 000 members, of whom some 40% are concerned with industrial marketing. There are 37 regional branches and eight industry groups, and a range of services and publications.

The British Institute of Management provides reading lists on aspects of marketing including industrial market research, and has these titles on its series of Management Check Lists:

List No.

5	Licensing agreements
16	Exporting – first steps
29/30	Launching new products
33	Physical distribution management
44	Marketing strategy and the advertising budget
45	Improving advertising effectiveness
54	Commissioning effective market research
65	Developing a corporate identity
74	Managing product design and development
75	Delivery on time
81/82	Know your customers/competitors

A typical marketing function, in a medium-to-large business, will be headed up by an Executive Director, who will be closely involved in determing the company's strategies.

The organisational arrangements in a high-technology company will usually put the Marketing Director alongside the Technical Director, with both providing their part of an agreed programme.

The Marketing Director will have responsibilities in these areas:

Product attributes

The user viewpoint on requirements, feedback of quality information, comparison with competitors in market standing, and liaison with his company's R&D

Product programmes

Promotion, trade shows etc., sales literature and sales organisation: the supply and distribution system, and the levels of service achieved: plans for each product line taking into account the probable life cycles

Input to company strategy

Marketing analysis, future scenarios, the external customer environment, identification of options, identification of potential competitors and of attractive market niches

The Chief Executive will hold the balance between the heads of the functions of marketing, engineering /R&D and manufacturing. The head of the finance function will tend to moderate their enthusiasms through the realistic financial appraisal of alternative courses of action. A company in high technology may have a Chief Executive who by background is an engineer, and possibly the founder of the business. He particularly needs the balancing influence of strong personalities as the heads of the other functions.

It is appropriate in some businesses to operate as Divisions, each serving a market sector: particular product ranges may be controlled by 'Product Managers', who draw on the other functions as necessary. Where the main consideration is geographical spread, Account Managers may service the needs of particular large customers or groups of customers.

Arrangements for export marketing need to be structured to suit the situation. In a steady on-going market, geographical branches are appropriate. But for large specialised plant, or special technology, the greatest impact will be by sending a team led by key technical people, under the administrative guidance of an export marketing specialist who knows the territory.

The experience of most companies in operating branches in other countries is that they should maximise the proportion of nationals on the staff. This is both more acceptable and more economical than manning with a high proportion of their own expatriates, who are more effective in the role of visiting advisers or trainers. Rather than start from scratch to establish a branch in another country, it may be easier to buy control of a small and compatible business as a nucleus. This requires a substantial investment, but brings a degree of dedication that may not be achieved by appointing Agents.

5.2 Market-driven management

A change in the market structure forces companies to rethink their strategies. They should be sufficiently in touch with potential developments, both political, governmental and technical, that they are not taken completely by surprise.

Such a situation has come about recently in the telecommunications industry. The Office of Telecommunications (OFTEL) was established in 1984 by Parliament, the purpose being to liberalise the telecommunication industry, with the aim of increasing the economic efficiency by bringing market forces to bear. Its operations have been outlined by the Director General, Professor B Carsberg (1986).

In parallel, the status of the Bell Telephone System in the USA has been changed, to introduce competition into what was a near monopoly. These Government-induced changes have come at a time when major changes in technology were being implemented, as part of long-term planning for completely digital networks, and a great increase in transmission of data, as outlined by O'Hara (1985). Another concurrent development has been the introduction of cellular telephone networks.

Viscount Caldecote (1986) has outlined the effect of changes in the international scene on the European telecommunications industry. Deregulation has removed the protected market previously enjoyed by manufacturers, within their national boundaries. There is the problem of a potentially large increase in imports, but an opportunity to invest in developments aimed at the international markets.

Western Europe has 38% of the free world's production of telecommunication equipment, but this has been declining, while that of Japan has risen to 15%. The Japanese efforts are well integrated, but those of the European manufacturers in the past have been fragmented. A first step will be for the European PTT's to adopt common technical standards and procurement policies.

While these changes in the market pose major problems of strategic decision for the major Groups in the industry, liberalisation allows smaller companies to achieve rapid growth in the supply of subsidiary equipment, subscribers' apparatus and specialist services.

Those in the industry are fortunate to be associated with a market having a growth rate several times that of the economy as a whole, but in their strategic planning they may have to make some calculated guesses of the directions that growth will take; similarly in broadcasting, satelite links, cable systems, and higher-definition television standards.

Clement-Jones (1986) has outlined 'intercept strategy', as a way of getting ahead and staying ahead in the technology race. This consists of identifying improved components or techniques, before they are generally available at economic cost. They are then designed into a future system, forecasting that the cost curve will fall sufficiently by the date of product release. He adds that it is safer to do this with only a few key elements, not changing everything at once: there should also be a fall-back technology available to bridge any period of delay.

Monds's (1984) book, in the IEE Management of Technology Series, explores the special features of the business of electronic product development, with some case histories. There are others in Chapter 6 of this book.

Gaythorpe (1984) contrasts his experiences as a user of electrical and electronic equipment in the UK and in a mining environment in Africa. He makes the point that the supplier must get to know the conditions of use in the market to which he sells, and must satisfy these three points:

- a well designed product using feedback from previous customers, to effect improvements
- the customer should have adequate resources for maintenance, both in men and materials
- it must be possible to operate and maintain the equipment within the technical infrastructure of its location

The marketing and sales organisation is the link that must provide this communication.

It is not only subtropical climates that have caused failures. Within the UK, electronic payphones have failed on damp days, television printed circuit boards have suffered surface tracking in the Scottish Highlands and Islands and, in the High Street, bank cash dispensers have failed because, in the rare hot summer, the security enclosures excessively contained the heat of the circuitry and mechanism.

Marketing is the interface with the user and his environment, and so should shoulder the primary responsibility for the specification of environmental needs of the particular market.

5.3 Global market

5.3.1 No longer a National market

In a review of Britain's export competitiveness, Lord Nelson of Stafford (1984) concluded that to recover its place in world markets, changes must be initiated in the areas of marketing, design, development and manufacture. He said that today no one in his right mind would invest in major programmes of product development without first identifying its world market.

Continuing, he commented that the Japanese have demonstrated how successful this can be, by the way they have progressively, over the last 25 years, identified and captured major sectors of the world market.

Beginning with hydroturbines in the 1950s, they progressed to ships, motor cars, motor cycles, electronic consumer goods and office equipment. The Europeans and the Japanese have been much quicker than us to exploit our historic export markets, and have in the process invaded our own home market – see Oakes (1987).

Further, no company can tackle everything at once, or every market at the same time, so the first need for management is to weigh up its own strengths and weaknesses. It must then establish a clear-cut policy of selection which takes these into account.

The concept of national markets is rapidly disappearing — there are now only world markets. Today every manufacturer must be customer oriented and develop a global strategy based on adequate market research. It was clear to him that improved marketing is essential to the future success of British industry.

5.3.2 The dynamic nature of markets

The virtues and weaknesses of the long reign of Queen Victoria in the 19th century were both the consequences of its stability. The Great Exhibition of 1851, at the Crystal Palace in London, marked a peak of achievement in industrial innovation. Britain had technology, which it was selling all over the world, to construct canals, railways, bridges, steamships and telegraphs, and the basic industries of steel, glass and textiles. Mass-produced components were used to construct the Crystal Palace, and in the dockyards for ship fittings. Already, the British windmills employed automatic controls, and programmed control of Jacquard looms by punched cards was well established.

The illusion of stability, at this peak of achievement, concealed from general perception the advances in other countries. One factor (leading to the American War of Independence) was that Britain forbade the export of the means of manufacturing goods. Like other attempts at sanctions, this did not work for long, because craftsmen emigrated with the know-how in their heads, and in recreating mills, machinery and processes in the new world, they dreamed up significant improvements. The process has continued in many other countries, at an accelerating rate.

A second factor was that new producers made for themselves. Distance isolated the producers, and transport was too costly for real competition, except for specialised and luxury products.

The third factor was the limited discretionary expenditure of the mass of population, after meeting the cost of their bare necessities.

The general trend in the first half of the twentieth century was for slow evolution, with a minimum disturbance from inflation. In the second half, the rate of change has accelerated, both economically and in technology. The mature countries have responded less rapidly than the New World and the developing countries.

Fig. 5.1 shows the growth of electrification is several countries, in terms of percentage of homes served. The UK reached the 60% point just after 1940, but, in terms of historic span, other countries were relatively close behind us:

Western Canada:		15 years
Costa Rica	:	35 years
Jordan	:	40 years

A fairly close correlation exists between the spread of electrification and the demand for industrial and domestic equipment. Also, there has been a general

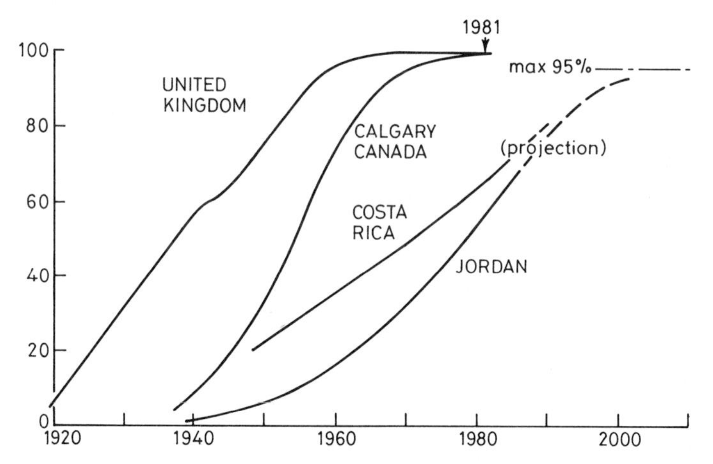

Fig. 5.1. *Examples of S curves, showing the progress of electrification in several countries, as the percentages of houses connected to the supply*

political preference for minimising restrictions on international free trade. Transportation cost is another constraint that has been drastically reduced.

When canals were constructed in the late 18th and early 19th centuries, they opened up inland commerce by making it possible for one man to move 50 tons of goods over 100 miles in two days, a hundred-fold reduction in cost when compared with the use of a packhorse or cart, or even Hannibal's elephants. This supplemented the earlier coastal- and river-navigations systems.

With the arrival of the railways, the bulk ton-mile productivity was improved by a further factor of 10, plus a factor of 10 advance in speed. Costs then stabilised for a century. Until the run down of Battersea Power Station at the end of the 1960s, it was more economic to send energy as coal from the North by sea, than by rail or high-voltage transmission.

Internationally, the introduction of standard containers in the 1960s offset the rising labour cost of physical handling, as transhipment of a 20 or 40 tonne load from ship to truck or rail takes only minutes. It became quite economical to manufacture consumer 'white goods' (refrigerators, washing machines etc.) in Italy, and ship them worldwide. The Japanese introduced purpose-built drive-on ships for bulk transport of new vehicles.

Air freight began to build up to significant tonnages about the same time, and became viable for lightweight and relatively high-value goods, such as exotic flowers and fruit, fashion textiles, photographic and electronic products. A freight charge of $1–5 per kilogram from USA or the Far East to Europe is viable for a fashion garment selling at $25 and weighing 100 g, or a camera worth $250 and weighing 500 g. The implications for stock control and inventory levels are considerable, as only a few days' requirements need to be carried by the importer.

It is now common practice to ship part-processed electronic assemblies and integrated circuits back and forth between continents before final completion.

These logistic dynamics concentrate large-quantity production to meet international needs, in the hands of the major manufacturing and merchandising Groups. But opportunities arise for many smaller companies to demonstrate their mobility, and to fill the gaps in the main networks of supply: for example, the 'silicon foundry' companies serve the needs of those who require custom chips.

5.3.3 Impact of international standards

The marketing function provides a company's interface with the international scene, and it has to have its 'ear to the ground'. Failure to realise early on that a new standard or practice is on its way may lose a company's place in the market.

A notable failure by the UK was to associate the availability of world wide container transhipment facilities with the potential demand for 'package' industrial plant in the developing countries, and for temporary and emergency applications.

One example is transportable gas-turbine generating plant, in the 15–20 MW range. If the components are shipped out and assembled on site as a conventional small power station, and then commissioned, with commissioning the process can take six months – if nothing goes wrong. This size of plant can be engineered ready-to-run within the framework of an ISO container, nominally 40 × 8 × 8 ft. This can be handled by ship, rail and road on transhipment equipment now available virtually everywhere in the world.

This was a US development, and the Japanese applied the same approach to 'packaged' telephone exchanges, supplied housed in the container ready for use. In this application the weights are low enough that an air-transportable version is possible, using a container housing of aluminium construction. The principle can be applied to industrial and chemical processing plant, so overcoming many of the difficulties outlined by Gaythorpe (1984), of limited local skills.

There is a need to feed information to design and development departments on the specific requirements of different countries. This was reviewed in a Colloquium (1983) on 'Design for Safety', which included contributions from BSI, ITT, Hoover and Austin Rover, and gave special attention to statutory requirements in various countries. There is slow progress towards international harmonisation of standards, but the next generation of products will have to provide for the current variations in requirements.

In industry there have been successes and failures over the years to secure international co-operation. When they succeed, it is the user who benefits, and demand increases. One early conceptual achievement was the acceptance by NTSC in the USA of a 'compatible' system of black-and-white/colour television, after prolonged controversy in the 1950s over other alternatives. This greatly eased the commercial and consumer transition, as users could change over to colour sets when it suited them to do so. The less satisfactory aspect was that, although Europe accepted the concept, different line standards and chroma systems were adopted.

It was not then forseen that there would be a huge market for domestic video cassettes, which consequently are incompatible between US and European standards. The manufacturers compounded the situation by introducing three different video formats, and now, with higher-performance tape, the Video 8 for amateur use.

Dominant manufacturers have from time to time set the standard that has been generally accepted. For example, IBM with the PC, Kodak with 'super 8' film and with disc cameras, Philips with the compact casette and compact disc, and British Telecom with Prestel.

Once again, television standards are in the international arena: where previously the NTSC compatible system was an elegant but simple waveform manipulation, the conflict between enhanced 625- or high-definition 1125-line systems is more difficult to resolve. In the intervening 35 years complex signal processing and conversion has become economically possible with relatively inexpensive chips. In a similar way, the problem of emission control of automobile engines may be resolved in the future by refined electronic control of the combustion conditions, rather than by purely chemical catalytic means.

A paper by Wheeldon (1986) reviewed the problems of identifying a product strategy, and outlines other 'case histories'; e.g. how the 10 in video disc missed the market window, because the video cassette got there first. The pay off from that investment in R&D was not lost, but postponed for some 3–5 years, when it provided the basis for the very popular 4 in compact disc, and for laser-disc data storage systems of multi-megabyte capacity, which may also be an electronic alternative method of publishing.

5.3.4 Assistance to smaller companies

From the foregoing it might appear that international marketing tends to be a monopoly of the large Groups. Despite this, a smaller company with a unique product or technology can usually form links in other countries. Filtronic, for example, (see Chapter 6 (6.2)) found that the potential market for its technology in the USA was ten times more than in the UK.

There is also the possibility of inwards licensing, for example, something available in USA but not yet marketed in Europe or elsewhere. The UK-based Rank Xerox organisation, with over 20 overseas subsidiary companies, began in this way in 1955, as Dessauer (1971) has described. The Haloid Company had in 1948 taken the rights to Chester Carlson's patents on electrostatic printing, but lacked the financial resources to market the newly developed Xerox machines beyond the USA. The J Arthur Rank Organisation already had worldwide offices for cinema film distribution and, in addition, manufactured precision optical and mechanical equipment; so it made an excellent match.

An outstanding example of outwards licensing is given by Ryan (1984). After a costly programme of R&D, Pilkington's perfected the 'float glass' process and decided to exploit it outside UK by licensing other companies. By 1972 they had 100 UK patents and corresponding patents in 50 other countries, the product of spending £7 m on R&D over seven years.

Their licensing and technical service income rose from £2.2 m in 1967 to a peak of £30 m in 1977. Although between 1981 and 1984 the main licenses ran out, receipts were still maintained at £30 m in 1984 and £26 m in 1985.

Potential overseas contacts can be identified through the large number of trade magazines, and by attendance at the specialist Trade Fairs. American companies are usually very willing to have preliminary talks with a visitor, on the basis that they too will benefit from an exchange of information.

Assistance is available from the Department of Trade & Industry, through the British Overseas Trade Board (BOTB). An extensive Statistics and Market Intelligence Library is located at the headquarters in Victoria Street, London, and there is a network of seven Regional Offices. The BOTB can obtain advice on individual markets through the commercial staff of more than 200 British Embassies and Consulates.

Both 'outward' and 'inward' missions are sponsored, and support is given to overseas Trade Fairs. The missions are arranged to encourage British exporters to visit overseas markets as organised parties, and to assist British companies to bring to the UK visiting parties of overseas businessmen and others who can influence exports.

Financial assistance for overseas visits is available to groups or individuals, to meet part of the cost of market surveys. This is known as the Export Marketing Research Scheme, and under it more than 600 grants are made annually. When an individual company applies, it must submit a detailed proposal and programme. It is sometimes feasible for several related but non-competitive firms to plan a joint mission, and group visits are often arranged by the main Trade Associations.

5.3.5 Desk research

Because of the cost in time and money to undertake an overseas market study tour, time invested in preliminary 'desk research' is well repaid.

Back files of the country's trade and technical journals are a valuable source, including the advertisements. The US Embassy Commercial Library is helpful, and other countries will have considerable information available at their Embassies, in particular their trade directories.

The libraries of the Engineering Institutions subscribe to some of the overseas technical journals, and a very wide range is held at the London Science Reference Library. The City Business Library by the Barbican and the Manchester Public Library are good sources of economic and business information. The Science Reference Library is in two locations: the section in Holborn (25 Southampton

Buildings) holds 12 500 titles of technical and trade periodicals from many countries, nearly 100 00 books, and over 20 million of the patents of all countries. The section in Aldwych (9 Kean Street, Drury Lane) holds literature on the life sciences, earth sciences and mathematics, with some 20 000 titles of periodicals. Both sections also have a large number of abstracting periodicals. The City Business Library has some 750 current periodicals, and specialises in Trade and Overseas Directories for over 150 countries, including more than 300 Directories from the USA. There are annual reports of 3600 UK companies, and extensive market data, and 100 newspapers from all over the world.

Principal banks with overseas branches usually have booklets giving the basic economic data on the countries they serve. The Institute of Directors Library in London holds a collection of these, and there is much information available at the Corby Management Information Centre of the British Institute of Management. The Department of Trade & Industry provides a Statistics and Market Intelligence Library at 1, Victoria Street, SW1, including foreign and UK statistics, market surveys and indexes, together with foreign trade directories.

If the market investigation relates to specific products or technologies, a database search is likely to turn up some useful articles or background information. The Technical Information Unit attached to the IEE/BCS Library is staffed with Information Officers who know their way through the many on-line data-base services now available (some 500 worldwide).

The depth of a search is most easily controlled if the inquirer sits with the Information Officer at the terminal. The service charge is of the order of £15 per hour to members. The search process at first throws up broad titles, most of which can be eliminated, and those selected printed out in greater detail. A recent inquiry in the field of 'clean room' technology produced long abstracts of detailed descriptions of installations at some 20 companies, mainly in the USA, and was completed in an hour. At Milton Keynes, The British Standards Institution maintains an extensive file and index of the standards of virtually all other countries, and queries of specific foreign requirements can be answered by searching the database.

Some other Institutions can provide search facilities; also some of the Research Associations, and the London and Manchester Business Schools. Services and specialised market reports are supplied by a number of commercial organisations, but their charges will be considerably higher.

A typical data-base service available directly to users is Micronet 800, which is accessed on a microcomputer via Prestel. The user can also have an interactive dialogue with an information centre. A typical search fee is £35, or £65 for a complete dossier of facts and figures from a variety of sources.

5.4 Product strategy, specification and development

5.4.1 Life cycles of products

The concept of a 'life cycle' in products was identified by Leavitt (1960), and developed further in books by Kotler (1980 and 1983). The typical pattern is the four-phase cycle, sketched in Fig. 5.2:

- introduction
- growth
- maturity
- obsolescence

If the product has genuine novel features and value, then in the words of R.W. Emerson in the mid 19th century:

> If a man write a better book, preach a better sermon, or make a better mouse-trap than his neighbour, tho' he build his house in the woods, the world will make a beaten path to his door

Thus, the initial sales take off may be rapid, but will die off when the market becomes saturated, or competition builds up. During the early period when the features are still unique, there is opportunity for considerable profit. When the profit curve eventually falls below zero, it makes sense to kill off the product.

However, this segment of business may be disposed of to another company, or

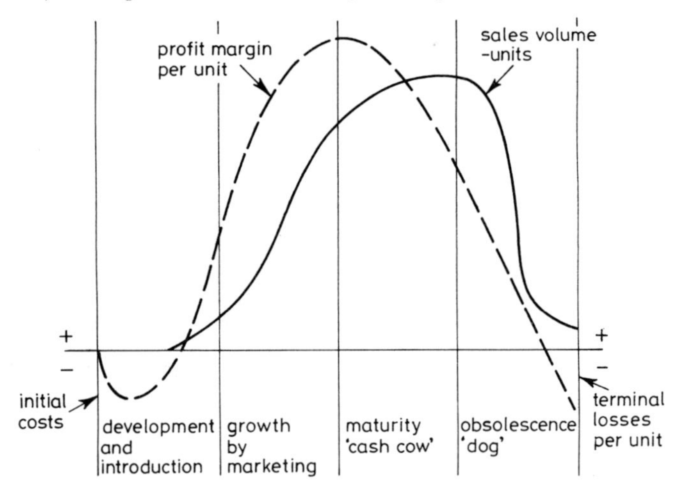

Fig. 5.2. *Simple representation of the product life-cycle concept (4-phase)*

by means of a staff buy out. The people who take over control may successfully turn it into a viable business again, by better marketing, by introducing improvements, adding value, and by cutting overheads.

The life-cycle model is more qualitative than quantitative, but it does help a management to appreciate where they are in the four phases: they must establish a balance in their product range between the new and the obsolescent.

Much that has been written on this subject comes from the field of consumer goods, initially soap and detergents. A little imagination is required when applying the concept to engineering products.

A balance of life cycles is secured by constructing a marketing or business 'portfolio', as described by Hedley (1977). This involves planning the programmes for products so that there is diversity between their phases. The company at any one time then has a balance of products in the obsolescence phase, and others being introduced, together with a solid core in between, to maintain earnings and net cash flow.

The idea was presented vividly in the 1970s by Hedley's Boston Consulting Group, positioning the four phases in an elementary matrix of (growth rate)/(market share), as in Fig. 5.3, the sectors being identified as follows:

Question marks: new products, growing rapidly, but requiring more cash input than they currently generate
Stars: leaders in their market, earning well, but demanding further investment if their market position is to be held
Cash cows: older products with a strong market position, low costs (through productivity improvements etc.), and needing minimal further investment in plant or R&D
Dogs: products with minimal growth, poor market share and low profitability

This categorisation has helped businesses to take decisions on a realistic basis, breaking away from a sentimental need to continue with the 'dogs'. It may be difficult to divest them, for example if there is a continuing obligation to provide

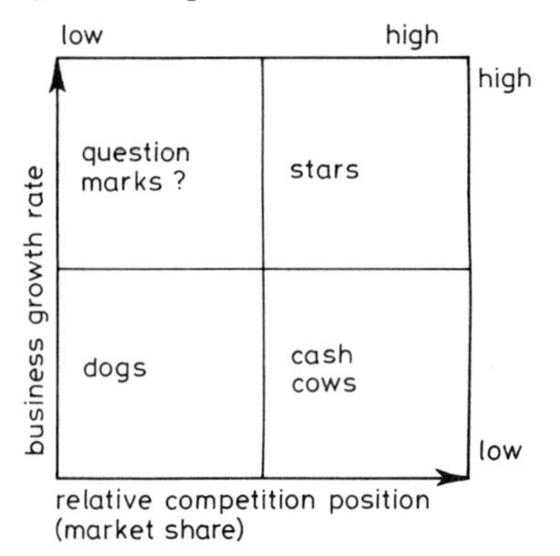

Fig. 5.3. ***The four phases of growth-rate/market-share***
NB. sketched here according to engineer's convention of *x* axis increasing to right

spares and service. One solution is to transfer the stock of spares (or their tooling) to an equipment maintenance company, which is able to operate on considerably lower overheads.

'Cash cows' are uninteresting, low-prestige products that a technically aggresive management may tend to disdain or ignore. A one-time 'high-tech' engineering development company developed a simple adjustable shelving system, and of all the products they worked on over many years, this is the only survivor, and a profitable one.

The decisions on 'loss-makers' have to be approached with care, because the basis of costing should be questioned. For example, if the same average overhead is applied to all products, then the potential 'cash-cows' are overcharged, for they make little demand on management, marketing or R&D resources: at first estimate what might appear to be a 'dog' may have real cash-generating potential.

Fig. 5.2 was first put forward in a discussion of supermarket goods, and over-simplifies the model for technical products, where 'introduction' is really two phases: 'development' and 'initial introduction' (or pilot production). Fig. 5.4 is from a paper by Wheeldon (1986), and shows five phases, with a heavy negative cash flow during the first two. It may take a relatively long time before the early

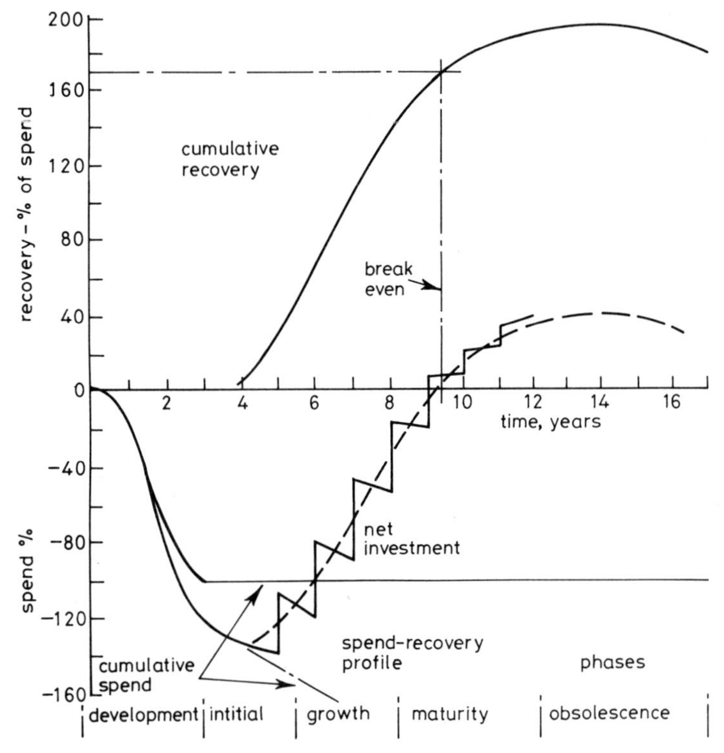

Fig. 5.4. *Product life-cycle (5 phase), with heavy initial investment in both development and introduction, making a total of 5 phases (from Wheeldon, 1986)*
In this example, break-even is not reached until fifth year from 'introduction' – the zig-zag 'spend' line represents annual outlay plus interest (sloping line) at 10% on net investment to date

outlay is offset by incoming gross profit, to achieve the break-even point. If the product is not withdrawn or replaced in the obsolescence (5th) phase, it will begin to make a loss, and reduce the peak of 'cumulative recovery'.

The sketch also illustrates the further point that capital charges on the negative cash flow must be recovered, and this further defers the achievement of break-even. The sloping part of the 'zig zag' in the diagram represents the accumulation of interest charges at a notional 10%. In the money market a return approaching

20% would be a realistic target: a straight-line outlay, continuing over five years, will amount to an extra 80% on the gross outlay.

This highlights the value of intensive product development: the capital charges are less for the same total spend, and the product is earlier in the market. The product development illustrated in Fig. 5.4 is something of a disaster, because the initial period of negative cash flow ($4\frac{1}{2}$ years) is nearly as long as the useful market life, and break even is achieved only very near the end of the period of marketability.

Hitchins (1986) has dealt with life cycles in complex systems engineering projects, where the obsolescence of 'mark X' can nearly overlap the launch of 'mark X + 2': his diagram reproduced as Fig. 5.5 shows the cycles in terms of manning requirements rather than cost. Foster (1986) stresses the advantage in taking the initiative with innovation early in the 'S-curve' cycle.

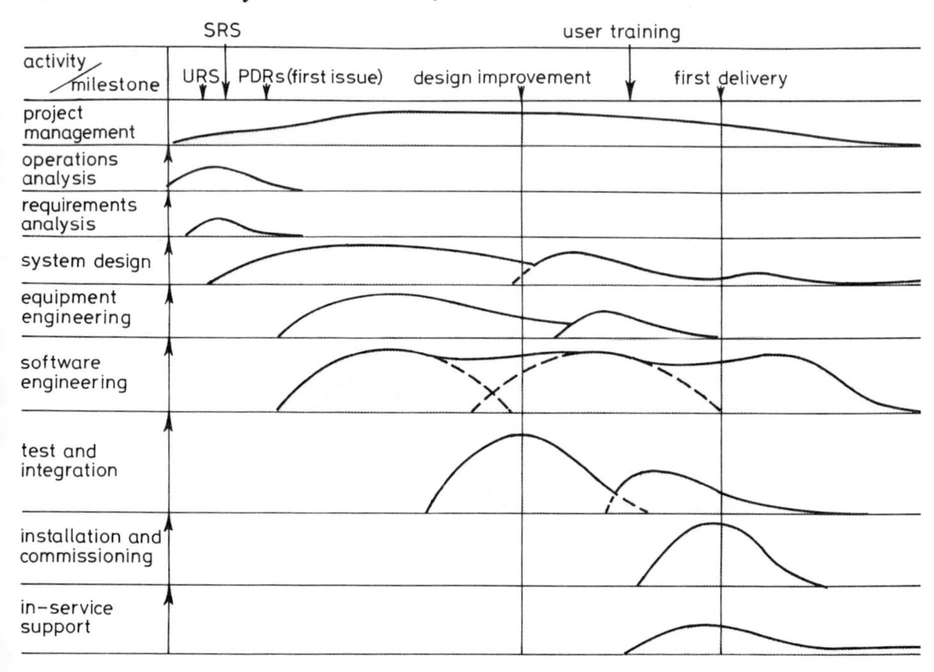

Fig. 5.5. ***Life cycle of a complex systems-engineering project in terms of resourcing required to man it (from Hitchins, 1986)***
SDR: performance and design requirements
SRS: system requirement specification
URS: user requirement specification

5.4.2 Engineering, research and development costs

A large industrial Group or holding company will usually act as 'banker' for its divisions or subsidiaries. It draws in their surplus cash flow, and then provides 'bridging loans' needed during periods of major expenditure on new development or other capital outlay. If the Group needs to borrow either short term or long term, it can get a better rate in the money market than a small company that has to go to its bank.

A sound business strategy is to ensure that, as far as possible, established products generate the cash flow for all new development, including a realistic allowance for those that fail. These costs are then written off in the current year and not capitalised.

It may be possible to develop a product or system as a series of modules or subsystems, each paying its way before the next is launched. This both smooths the cash flow and reduces the technological risks: several small steps instead of one major new system.

The importance of an integrated approach to market needs and engineering requirements is illustrated in a pair of papers from Plessey Company authors. T.S. McLeod (1983) described how it was recognised that computer aids to design, manufacture and testing were necessary to improve the company's engineering productivity.

A major study project was set up for the choice and implementation of a company-wide CAD system, to meet the needs of end users in the market segments of telecommunications, radar, aerospace and data handling. The techniques involved were testgear, custom IC design, mechanical design and PCB layout. McLeod shows how the project manager co-ordinated the inputs and contributory activities of four trading divisions, four technology working groups, and four implementation task teams.

The other paper, by A.W. Jones (1986), shows the edge that advanced technology has provided in the international market place for radar equipment. A family of radars was developed over a period of years, for a variety of clients and end uses. The systems were structured on a modular basis so that specific improvements could be introduced in stages, without re-engineering the whole system.

An important contribution was to forecast the downward cost trends of key components, and so determine when it would be cost effective to phase them into the commercial product. For the signal-processing functions, LSI chips were designed on the 'megacel' CAD system, which engineers can learn to use very quickly. It simulates both the logic and the dynamic performance of the chip, and identifies the errors that have to be corrected. Jones comments that the international radar market is very competitive, and the 'name of the game' is careful management of expenditure on technology, at just the right time to meet market needs.

5.4.3 Strategies and market niches

Even the largest Groups can no longer cover every sector of their market, but have to follow a selective strategy. It is important to identify when a line of products moves from being a 'cash cow' to the category of 'dog'. As already mentioned, the costing conventions adopted should be critically examined, as a 'dog' hived off as a separate commercial unit may revive and produce a modest profit and a return on the capital tied up in it. Such routine operations require a style of management different from that of a new innovative product, but more akin to the modest enterprises that comprise the bulk of the national economy.

Top management should take a synoptic view of their company's total market environment, which for most in technology means a global view. They will identify potential opportunities and threats, and must balance them against their company's strengths and weaknesses. It is not feasible to tackle the market on all fronts simultaneously: the options should be ranked systematically on the principle of the 'top ten', which are then investigated in greater depth, both outward-looking and inward-looking.

Market research and technical intelligence may identify a 'market niche' that is opening up, but with only weak or negligible competition. A company will possess a bank of 'know how' from previous work, which may give it a unique advantage, if the right applications are spotted. Then it must follow a strategy of going for it boldly, ahead of potential competition.

Conversely, having pioneered a new technology, process or product, if this should grow to require resources far beyond those of the initiating company, with profit margins falling as volume rises, the original company is well advised to make a planned withdrawal from the market, rather than sink resources into a losing fight.

This is a difficult decision for a technology-based top management, who previously have been personally and emotionally involved in the original development. It is easier to take if they can look forward to future possibilities, rather than backwards to opportunities that have passed.

A useful concept is that of 'technology push' versus 'market pull'. The traditional stereotype of an inventor was someone with an idea, for which he was trying to find a market. Sir Frank Whittle had a long struggle before the idea of the jet engine received substantial backing, and Sir Christopher Cockerell, a research engineer with Marconi and amateur yachtsman, had a similar battle to gain commercial acceptance of the hovercraft principle.

It is in the academic tradition to further the body of knowledge through our system of learned societies and basic research, but the processes are now so expensive that any commercial organisation has to confine itself to applied research in a narrow sector. The broader programmes now have to be co-ordinated on the lines of the Japanese MITI organisation, the Alvey Directorate and EEC's Esprit.

'Market pull' works the other way round – a company identifies a market niche that it is well qualified to serve, and makes a wholehearted effort to succeed in it. To spot such opportunities and 'market windows', requires first class 'upwards communication' from those in touch with the details, to top management who have to make the major commitment of resources. A half-hearted attempt to occupy a potential niche is likely to end in wasted resources; too little, too late.

5.4.4 Scenarios, and commitment to change

In an environment of changing markets and changing technology, the successful companies are those that think and act before the changes take full effect, rather than react to them later.

A simple matrix illustrates the point (Fig. 5.6). Corporate strategy must strike a balance between the potential profit or risks of a new product in a new market, and the poor returns and prospects of an old/old product. A balanced portfolio of products will maintain a more steady year-by-year progression. The bets have to be placed well before it becomes clear which will turn out to be 'cows' and 'dogs'.

A very selective approach is necessary towards new ventures, as the most risky are those in the new/new category. They need to be balanced by core business in the old/new and new/old areas.

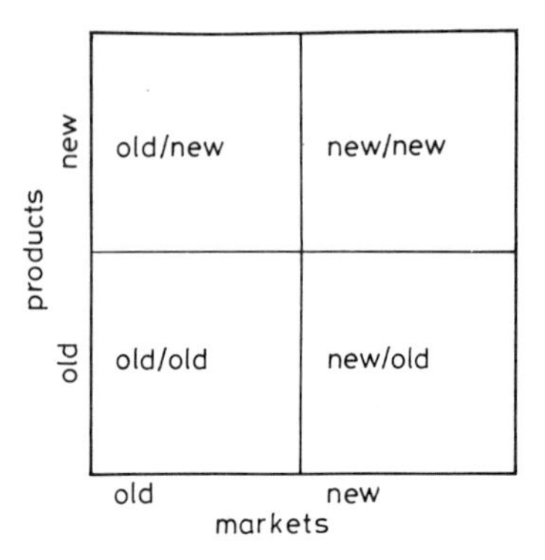

Fig. 5.6. *A balanced portfolio of products will balance profits and risks, for the options shown in the simple matrix of old/new products and old/new markets*

Scenarios are a vehicle for management mind stretching, and a means of exploring a wide range of future options with a minimum risk and expenditure. They can be prepared quite cheaply at the level of 'desk research'. Alternatively, major project proposals can be taken through all the stages of preliminary design, but then put on 'hold' if the market is not right at the time when major capital commitments have to be made. Limited development can be allowed to continue, in order that particular features may be demonstrated. Verschuur (1984) goes into much more detail on technology opportunity forecasting, in his sections 2.3 onwards.

An advantage of the 'scenario' approach is that engineers, designers, marketing people and management are all involved in a team effort. There is a mandate for wide-range imaginative thinking along alternative routes, without first having to secure overall approval and commitment. This comes later, as the knowledge increases and the choices are narrowed down. Scenario reviews counteract the 'tunnel vision' that can develop in a team working too long towards one bright idea or single solution of a problem.

The trend is towards vertical integration in markets, rather than a blanket coverage of 'everything electrical', which was once the motto of a well known Group. Attention in detail to a particular niche market provides a concentration of effort, and opportunity for direct contact and acquaintance with the users, appreciation of their needs, and feedback of their experience with the products.

A useful way to assess position in a complex market is a morphological chart, e.g. the example in Fig. 1.13. This illustrates the relative performance of different types of computer printer, on cost/performance axes. This type of diagram is a quick approximate guide to market comparisons, and the identification of gaps in the available products.

In the USA, General Electric (1986) has for some years followed a strategy of reducing diversification, withdrawing from certain market sectors, in order to concentrate efforts in selected market segments. Over this period it has discontinued its business in computers, domestic appliances and other traditional products.

The Philips Group (1986) is now the only international corporation to match competition from the Pacific Basin in consumer electronics and other electrical products, an area largely abandoned by the major US manufacturers. It has strengthened its position with a number of joint ventures with Japan and other countries. It has forecast that in 5–7 years more than half of sales will be in products not yet in existence.

5.4.5 Proposals, tenders and negotiation

Serving a global market introduces currency risks, both in launching products, and in tendering for major contracts. During the time interval between the initial planning, and the cash flow on delivery or completion, there may be large fluctuations in the rate of exchange.

For example, the value of the £sterling in $US fell from a high point of 2.45 in late 1980, to a low 1.05 in early 1985, and then back above 1.60 in early 1987.

The position is more complex than this, because there is variable linkage with other key currencies, such as the yen and D mark. As explained in Section 3.7.2, forward options can be taken on the money market, but only for a relatively short period ahead, relative to the 3–5-year periods of product life or that of a major contract. This is discussed more fully in papers by Babbel (1982) and by Boothe (1985). International currency fluctuations are responses to both economic and political factors, and should form an element of any scenario.

Major constructions projects in the multi-£million class require a substantial investment in preliminary work on the proposal, whether for competitive tender, or as an in-house project of a large organisation. Francis (1986) describes the approach to major capital projects in British Petroleum plc. A clear 'project definition' is necessary to avoid waste of time and money later. As this technical definition progresses, so does the accuracy of the cost estimate. BP employ a phased approval procedure, so there is a critical re-examination from phase to phase. It has been found that time and money is saved by spending more on the early stages of definition and design. The book by Sassoon (1982) describes the special requirements of tendering when the financing is by international finance agencies.

In construction work, tendering is simplified by the existence of 'standard forms of contracts', and with competitive tendering there is considerable scope for exploring alternative technical solutions, as the client's original stated requirement may not be the most cost-effective way of meeting the requirement. A recent interesting example of this was the additional Thames crossing link for the M25 motorway, where the estimated costs were of the same order of magnitude for the alternative engineering solutions of high-level bridge, bored tunnel or sunken-section tunnel.

In large international bids there is usually an invitation to make 'pre-qualification' submissions, open to all. Then a short list is selected, and serious discussions begin on details. The combined costs of the bidders is high, as possibly ten may be involved, but they may strengthen their positions by coming together in consortia. The winner of the tender has usually provided a strong technical team spending months on site, with pre-agreed negotiating limits, plus offers of special finance arrangements, and a great deal of patience.

Scott (1985), and also Nierenberg (1973), point out that in successful negotiation everyone wins, in contrast with an adverserial win–lose attitude, and there is a considerable legal and psychological content. Other cultures than the Anglo-American put much weight on the quality of the personal relationships formed between the parties in negotiation.

5.4.6 Quality systems and product liability

A manufacturer or a design consultancy has a responsibility for the quality of their work, which they must exercise internally in the corporate sense. There are rules of professional conduct that apply to individuals of particular Institutions, of which the Professional Brief of the IEE (1982) is representative. The client will have the expectation of assurance that he can depend on the quality controls exercised by the vendor, and he will have the support of contract law. Where safety regulations or other legislation applies, there is also the backing of criminal law.

The British Standard Institution's 'Quality System', BS5750, evolved from the Ministry of Defence MoD 05-20 series, and the related NATO standards. Its development during the last decade has been under the guidance of Derek Spickernell (1983), formerly Director General of Quality Assurance in the MoD, subsequently Director General of the BSI, and then Vice-President of the International Organisation for Standardisation (ISO), together with his successor as Director General at BSI, Ivan Dunstan. The British quality management system has been adopted as the basis for the international standard, ISO 9000. The individual national standard institutions will continue to draw up detailed technical standards for particular products.

This family of standards is fundamentally different from the traditional form of specific and detailed standards. The requirements for a system of quality management are specified in general terms: the user must then interpret these concepts in terms of his own operations, and ensure that the requirements are fulfilled by his own procedures. He is expected to document these procedures (usually in the form of a 'Quality System Manual'), and comply with them by self-control. Allender (1984) has outlined the process.

When a company applies for certification by BSI, it will first be assessed by BSI itself, or another approved third party such as Lloyds Register. A company is accepted after any deficiencies have been rectified, and there are now some 10 000 on the Register, about twice the original list of MoD approved firms.

The assessors are concerned not only that an adequate system of quality management exists, but that it is understood by staff at all levels. An important feature of implementation is the motivation of all personnel to make their individual contributions.

It is in connection with staff motivation that 'quality circles' make a major contribution, as described by Robson (1982): usually they are coordinated by someone trained in the role of 'facilitator'. The 'circles' provide opportunities to involve all workpeople, but they must be integrated with the overall quality management system. It is usual to identify an appropriate person in a company as 'Quality Manager', although in smaller firms this function may be paired with another.

A purchaser will specify that his requirements should conform to BS5750, and he will not expect to have to undertake any incoming inspection, other than random spot checks. This is a useful saving, together with the other feature that 'multiple certification' is eliminated. Previously, a large supplier would have to receive assessment teams from a number of client organisations, which was a time-consuming process for both parties.

Mann (1984) has described the role that a consulting engineer can play in applying the quality assurance concept in a project programme, on behalf of his client. The system as described in BS5750 covers every function of an organisation in general terms. This has to be interpreted into the language of the particular industry, in preparing a manual, with the procedures defined in a Quality Plan. This can be done effectively in such diverse situations as mechanical engineering, food processing, metallurgical processes, textiles and electronics manufacture. There is also a simplified scheme covering stockists and distributors.

Encouraged by Government, major public procurement agencies specify these requirements to their suppliers, for example, Lomas (1983) has outlined the arrangements in British Telecom. The Institute of Purchasing & Supply has assisted its members in the private sector to adopt the system. There is a professional body for those personally involved, the Institute of Quality Assurance. The IEE (1984) has issued a Professional Brief on 'Product Liability', which summarises some of the other statutory requirements.

The basic principles of the system apply equally well to the provision of services; and also in the preparation of software, where the British Computer Society has been active in developing quality assurance. The BSI and the Design Council are collaborating in the development of a British Standard for managing product design (Constable, 1985), which will be based on the quality management concept.

5.5 The market for services

5.5.1 The service industries

The balance between numbers employed in manufacturing and in the service industries has changed considerably in recent years, mainly through three factors:

- manufacturing has become less labour-intensive, through new processes (moulding etc.), mechanisation and automation

- a higher proportion of goods are now imported
- maintenance of higher environmental standards, and increased discretionary expenditure

It is sometimes said that 'service' is contrary to the mentality of some nationalities, perhaps through subconscious linkage with 'servility'. Peter Drucker (1973) has done as much as anyone to put into perspective the contribution of the non-manufacturing institutions, banking, retail, public utilities, medical services and professional services. Whether private enterprise, or instruments of government, these institutions exist to meet clients' needs.

In the public sector an outstanding early model was the Tennessee Valley Authority (TVA), the big regional electric-power and irrigation project in the south-western United States, started in the 1930s. While basically an engineering project to construct dams and generating facilities, the Administration did a great deal to enrich the infrastructure, with low-cost rural electrification as the catalyst, a pattern that has been followed since in the developing countries of the world. The motto of the Public Utility of Costa Rica in Central America (ICE) is 'electricidad – fuente de prosperidad' – the source of prosperity.

Until recently the Bell Telephone System (AT&T) in the USA was a virtual monopoly. However, it had been the leader among service institutions from the early part of the century. This came about because the AT&T President, Theodore N. Vail, defined the company's mission as 'our business is service'. He appreciated that he needed community support, as the company was politically susceptible to nationalisation.

For more than half a century, the recurrent theme is marketing, advertising and staff training was friendly, prompt and efficient service – a surprising experience for visitors from most other countries.

This definition of 'what is our business' was backed up by a management system that monitored and measured performance, with a series of 'standards of service'. A similar system was introduced in the UK when the General Post Office became a public corporation in the late 1960s, and was then divided into the Post Office and British Telecom.

All the UK Public Utilities now have their own Codes of Practice, defining the services that the public can expect from them. The Codes are agreed in conjunction with the National Users Councils, and the Office of Fair Trading.

Standards of service, based on measurements of performance, are the basis of many successful businesses, particularly those where there are many branches, and where uniform high standards are desirable. Obvious examples are the car rental chains, hotels and fast-food outlets, where staff are trained to a high standard for relatively straightforward tasks, and motivated to put the customer first.

The same approach has been found essential in the marketing of high-technology products and services. Large and expensive machines in factories and the construction industries have a high hourly cost if they breakdown, and even higher secondary costs. Computers and office machinery require to be backed up with highly competent maintenance services.

There are technical contributions to enhanced service; for example, the liberalisation of the telephone service and the availability of standard connectors has greatly simplified the 'installation' of a variety of equipment.

The whole trend in electronics, in computer peripherals, and in electromechanical and hydraulic systems is towards 'black boxes' that provide a function and are readily interchangeable, but do not require diagnosis by the users of the internal workings.

Monitoring by management of a service operation is essential. In some companies the top executives serve at the point of customer contact from time to time; so they retain a feel for the business. Training stresses that every employee, down to telephone operator and delivery man, represents the organisation, and can enhance or mar the client's image. One bad experience can wipe out a long programme of effective public relations. A system of documenting and analysing complaints is essential, plus feedback of corrective action.

Two books may be mentioned in this field, McCafferty (1980) and Albrecht (1985). The latter comments that some engineering people act as if the organisation exists to support their intellectual hobbies. Top management has to build a 'service minded culture' throughout the organisation. It can be helpful to analyse the experiences of customers. British Airways did this, and found that travellers responded to four key factors:

- care and concern experienced
- spontaneity of staff
- skill in overcoming problems
- recovery

This last factor was newly identified by the survey, and assessed how resourceful were the staff when something went wrong, and exceptional action was necessary. This is something that is especially encouraged within IBM, and perhaps is an echo of the legend that Rolls Royce would send an engineer to anywhere in the world where an owner was stranded.

5.5.2 Consultancy services

A growing market exists for consultancy services, partly due to the increase in specialisation and also because there is a move away from employing large headquarters functional staffs. It is easier to call in the specialists for limited assignments and projects. This makes for a sharper and more specific approach, as the task will be clearly defined in terms of reference. The internal staff man tends to march to a slower drum.

A significant proportion of chartered engineers serve as consultants, and much of this is of the nature of project design and management. Fig. 5.7, based on data provided by the Association of Consulting Engineers, shows the considerable growth of international contracts under management. Directories are published by the British Consultants Bureau, and also by the Association of Consulting Scientists.

A survey of the work of engineering consultants edited by John (1984) contains 15 contributions on various aspects including scientific and management consultancy. The latter is described by Tisdall (1982), membership being about equally divided between those experienced in marketing, engineering, personnel and accountancy.

It is estimated that some 10 000 chartered engineers are engaged in engineering consultancy (some as principals), and that in all disciplines of management consultancy there are some 5000 professionals.

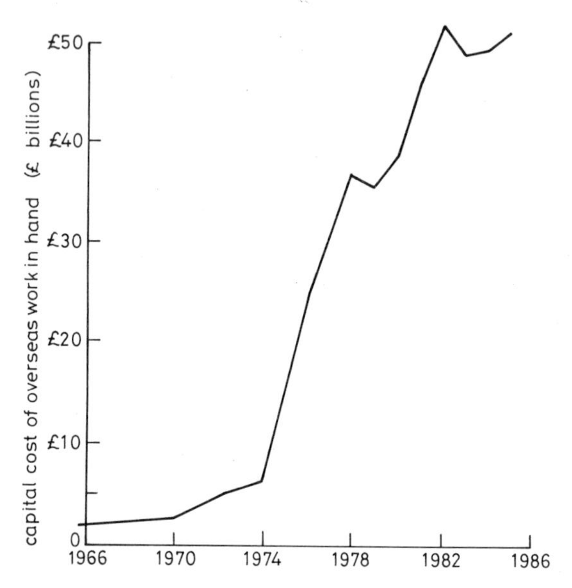

Fig. 5.7. *The work of consulting engineers (from Association of Consulting Engineers)* Total value of international contract currently under management (contracts may continue over several years)

The consultant, employed for a limited period, can assist in three ways:

- to diagnose, and recommend action
- to implement change and, where necessary, train and motivate those involved
- to be an extra pair of hands in a peak period

A booklet on 'How to select a management consultant' is jointly issued by the British Institute of Management and the Institute of Management Consultants.

5.5.3 Self-employed status

Experienced engineers have the option of 'going independent', and there is considerable scope for their services amongst clients who have no need to employ permanently an internal engineering team or senior engineer. The Professional Brief on 'Consultancy' issued by the IEE (1984) advises on the business aspects of taking this step. This was prepared in conjunction with the Royal Society of Chemistry and the Institution of Mechanical Engineers, and it is noted that the I.Mech.E. runs a course for those contemplating setting up as a consultant.

A high proportion of computer-systems specialists now work on a freelance basis. This is particularly appropriate in this field, as a client will implement a new system from time to time, but does not need to employ staff permanently who can work at the conceptual level. The freelance can work under the umbrella of a Contract House, or put his name on Registers and build up personal contacts. The British Computer Society advises on professional freelancing, and has a scheme for assisting the professional development of its members. The Institute of Management Consultants provides training seminars: it recommends those whose aim is to become independent consultants, to first get at least three years experience on the staff of one of the larger firms.

'Networking' was a concept introduced by Rank Xerox in 1981, and is described by Drew (1985) in 'Networking in Organisations': employees were encouraged to set up as independent contractors, and were assured an initial amount of work, while they built up other contacts. This goes a long way towards overcoming the main problem of an 'independent': that much time is spent self-marketing. Tax and expense allowances when working from a home base are relatively favourable, but it is advisable to put fairly substantial sums aside through a suitable personal pension scheme. The levels of allowance change from one budget to the next, and the recent trend has been to favour the self-employed.

Service abroad is a very worthwhile experience, but unfortunately the incidence of UK taxation is a nonlinear function. To be tax exempt, a full year's absence is necessary.

5.6 Working abroad effectively

A Professional Brief on 'Working Overseas' is available from the IEE (1985), and provides a fairly comprehensive checklist of what has to be considered in connection with conditions of employment, living conditions overseas, and medical advice. The time to negotiate adequate conditions and allowances is before leaving UK, and where possible advice should be sought from those who have visited the territory. As already mentioned, for tax reasons visits overseas tend to be either brief or lengthy.

Although brief business trips to a strange country may often have to be undertaken at short notice, time should be found to secure some briefing, obtain maps and other background data, as already outlined in Section 5.3. The incidence of 'jet lag' varies greatly between individuals. It does help on long flight to take plenty of non-alcoholic drink, moderate amounts of food, forego the movie, and practise Yoga. There is a cult of rushing out and rushing back, but it does a lot of good to the person, and little harm to the business, to include a weekend or a few days break. The business consequences of 'being under the weather' are much more serious than the delay of a day or two. The business tour of several weeks can be very exacting, if attempted to a tight schedule.

For longer assignments in one location, it is wise to find time to get some knowledge of history, politics and geography. The Embassies in the UK are usually helpful, and a good bookshop or Public Library can usually turn up some literature on the country.

'Cultural shock' is a potential problem that affects some more than others. If the new location is completely strange, a sense of disorientation sets in, and it is not uncommon for people to return quite quickly to the UK with a medical certificate for 'psychosomatic disturbance'. A positive attitude and interest in the new location is an excellent antidote.

Those one meets professionally and socially at an overseas location respond very well when the visitor is seen to have a genuine interest in their country and culture. Preparatory courses, preferably for both husband and wife, are provided by The Centre for International Briefing, Farnham Castle, Surrey.

Languages can be a problem for those who have not had previous opportunities for practice. It is recommended to take an intensive course before leaving UK, but this can occupy two weeks full time: such courses usually concentrate on the spoken word.

Once on site, the services of a bilingual secretary can be requested. It is then wise to do all documentation in both languages, either as checklists, keyword lists, or with text 'exploded' into main, secondary and tertiary headings. If the versions in both languages are typed out in similar style and pagination, it is very easy to follow, on the English version, what is being gone through in the second language. The Kompas Trade Directories are indexed in several languages, and this provides an up-to-date guide to usage in technical terminology of materials, products and services.

While depending on a bilingual counterpart for the formal material, it is much appreciated if some verbal attempts are made to use the language of your hosts. This can be done by rehearsing some social conversation that can be used for 'openers'.

In the short term, effort put into verbal facility is more rewarding than bookwork. A reasonable pronunciation is very necessary, and can be acquired by repetitive practice with a pair of cassette recorders, one to play a language-course tape and the other to record your attempts to replicate it.

For a prolonged stay, it is useful to arrange for individual coaching. The British Embassy can usually recommend someone, perhaps attached to the University. Conversational practice can also provide a good deal of cultural insight that otherwise might be missed.

Traditional two-year overseas contracts are a hangover from the days when the outward boat trip might take several weeks. The adjustment problems outlined can be greater if the family accompanies the expatriate: if surmounted, it can be a unique experience. As Rudyard Kipling wrote:

> and what should they know of England who only England know

5.7 Reading list

BAYLISS, J.S. (1985): 'Marketing for engineers' (London, IEE/Peter Peregrinus) 391 pp.

BAKER, M.J. *et al.* (1983): 'Market development: a comprehensive survey' (Harmondsworth, Penguin) 240 pp.

COWELL, D.W. (1984): 'Marketing of services' (London, Heinemann/Inst. Marketing) 340 pp.

CROUCH, S. (1980): 'Marketing research for managers' (London, Heinemann/Pan).

HART, N. (Ed.) (1984): 'The marketing of industrial products (London, McGraw Hill, 2nd edn.) 212 pp.

KOTLER, P. (1983): 'Principles of marketing' (Englewood Cliffs, NJ.: Prentice Hall, 2nd edn.) 676 pp.

MORSE, S. (1982): 'Management skills in marketing' (Maidenhead, McGraw Hill) 150 pp.

RYAN, C.G. (1984): 'The marketing of technology' (London, IEE/Peter Peregrinus) 144 pp.

VERSHUUR, J.J. (1984): 'Technologies and markets' (London, IEE/Peter Peregrinus) 212 pp.

WILLS, G., CHEESE, J., KENNEDY, S. and RUSHTON, A. (1980): 'Introducing marketing' (London, Heinemann/Pan, rev. edn.).

Chapter 6

New ventures

6.1 Summary: the pattern is no pattern

New ventures are now looked upon very favourably, and considerable assistance is available to an entrepreneur, both financial from the City, and from facilities in particular localities. But it will take a good deal of his time and energy to pursue the business side of his venture, just in the phase when he may also be engaged in intensive technical development of his product.

It was a study by David L Birch of MIT (1979), that first quantified in the USA the role of small businesses in job creation. His analysis showed that enterprises with less than 20 staff created two-thirds of the new jobs in the 1970s, and the proportion has increased since. This swing is in part due to the search by large organisations for economies of scale and reduction of overheads, so that they have tended to reduce their employment.

In recent years the service industries have been the main area of growth both in the USA and Europe. We have to be careful about drawing particular conclusions, as the proportion of new businesses based on technology is relatively small, and they have special characteristics: if successful they have a growth rate higher than average.

There is a unique set of rules for initiating high-technology enterprises – the success stories are diverse in their characteristics, but these are contributory factors:

- being market-led (rather than promoting a pet invention)
- flexibility in revising objectives
- concentration on a clear-cut market niche
- excellence in this chosen field
- a compatible team
- optimum growth rate (steady, but not spectacular)

At one extreme, there is the example of Hewlett and Packard, who, from a part-time activity in the home garage in 1938, built up to a turnover of more than $6 billion and over 80 000 staff. At the other end of the spectrum are many failed enterprises, and, in the middle, a large number working in a limited market segment that they have either identified or stumbled upon. They provide a product or service, and continue to exploit it modestly.

There are several hurdles at which an enterprise can stumble or fail:

- failing to identify a genuine market need
- product or service lacking uniqueness, or inadequate
- insufficient finance during start-up and early growth
- failure to keep cash-flow balances under control
- the need to increase market penetration
- no follow-through with further products/diversification
- too many products, too early
- failure to identify and match competition
- instability and inadequacy in the team

All these elements are susceptible to quantification, both in magnitude and in rate of change, and people with scientific and engineering backgrounds are better qualified than most to get a feel of their situation. If they fail in this it is because they are too busy, and manage inadequately their working time.

A specialised product or service, directed to a professional market, has more limited growth potential than an innovation in consumer goods, but it is less likely to attract fierce competition. Generally speaking, it is less attractive to pursue growth of volume than to maximise profitability on a moderate turnover. But there will be exceptions to any generalisation: it may pay to exploit a unique opportunity to the full for a short period, then withdraw or sell out as competition builds up.

This illustrates the advantage in mobility of the entrepreneur, by comparison with the large Groups. Conversely, the latter may prefer to invest less in their own product development, bide their time, and then buy up appropriate entrepreneurial firms. They often fail to retain the founder management, who prefer to take their capital gains and apply them to further ventures.

The Fig. 6.1 summarises a few case histories, and their various characteristics. We are grateful to those involved for permitting publication. These are successes, and understandably those connected with failures are less forthcoming. If there is one secret for survival and success, it is to achieve profitability, on however modest a scale, as early as possible. A business with potential has no difficulty in raising funds for expansion if it can demonstrate that it is competently and tightly run, and has already a record of consistent profit and growth.

It is a basic economic phenomenon that innovative products can attract a higher profit margin than basic commodities. This profit ratio can sometimes be high enough to achieve self-financing, if the founders are willing to take very little out of the business at first. Where manufacturing is involved, injection of money is almost always essential to finance equipment, stocks and creditors during growth.

As explained in Chapter 3, venture capitalists will assist in the expectation of a long-term return, taken as a share in the capital value of the company. The banks are not in the risk business, and will only provide short-term finance in return for interest and instalments of loan repayment, and will require some form of security: often this has to be the entrepreneur's house. Both sources will put much weight on their assessment of the personal calibre and past successes of the principals in the enterprise.

Although flexibility is desirable, in order to survive by adapting to the market environment, difficulties can develop in the organisation structure and ownership of an enterprise that diverges form its initial concept. The originators often start up in the spirit of a partnership or co-operative, and later have to rearrange this initial basis into a more conventional structure of Chief Executive, management team, employees and shareholders.

The chart in Fig. 6.1 summarises the characteristics of the five enterprises whose case histories are outlined in the following Sections 6.2–6.6. In each instance their success has come from an ability to be flexible and to adapt to changing circumstances and market needs. These 'snapshots' were taken early in 1986, and by the time the reader reaches this part of the book, the enterprises may themselves have entered a new chapter.

6.2 Case History: Filtronic Components Ltd.
(Shipley, Yorkshire and San Diego, California)

This example draws on a lecture given by Professor J. D. Rhodes, and published in the *IEE Proceedings**. In his paper he identified the value of smaller firms as technical innovators, and their place in the national economic scene. Less than 1% of newly formed companies are concerned with technology, and even fewer are in 'new technology'.

He recommends securing the advice and assistance of a lawyer and an accountant from the beginning, and to take care to establish a good 'track record' with suppliers, as business references can be quite important in the early stages. Suppliers will expect prompt payment, but customers may be slow to settle their accounts; control of cash flow is a key function.

Professor Rhodes learned the hard way, having founded this company in 1977 as a part-time activity, while occupying his chair at Leeds University. The basis of the innovation was to employ multilayer printed-circuit constructions to produce microwave components, rather than to fabricate from solid metal.

The initial products were microwave filters with a suspended substrate stripline, for use in the front end of radar warning receivers. More recently complete receivers have been developed, using precision printed circuits.

The first full-time employee was engaged in late 1979, and in the summer of 1981 the firm moved to large premises in Shipley. Late that year they secured the support of a new Edinburgh-based venture capital group, Advent Technology,

* RHODES, J. D.: 'Setting up small high-technology companies' *IEE Proc.* 1983, 130, Pt. A, pp. 398–400, and case study in 'Opportunities in management for professional engineers' (1988), London: Peter Peregrinus (to be published).

initially ○
currently ●
minor ◐

		FILTRONIC COMPONENTS	MICROSYSTEMS DESIGN	BARKWAY ELECTRONICS	MILLENNIUM	S I D Inc.
ORIGIN OF VENTURE	start-up	●	●			
	spin-off				●	○
	revival			○		
FINANCE SOURCES	self-financed	○			●	●
	Bank loans	●	●	●	●	
	private loans/invest			●		
	Merchant Bank	●	●			
	Venture Capitalist	●		●		
	U S M		●			
	Stock Exchange List					
ACCOMODATION	'home'				○	○
	rented				●	
	long lease	●	●	●		
	bought/built				●	●
INTERNATIONAL	Agents			●		
	direct sales	●	●		●	
	linked companies		●			
	owned companies	●				
	UK based only					USA base
PRODUCT/SERVICE	consultancy	○	○		●	◐
	systems/software		●		●	●
	prototypes	●			●	
	turnkey applications				●	●
	own-name products	●	●	●	◐	●
	supply OEM's	●	●		●	
ASSOCIATED COMPANIES	none			●	●	
	related fields	●	●			
	diversified					
	part of large Group					

Fig. 6.1 *New ventures: comparison of case histories*

which is backed by Scottish insurance companies. AT took a minority stake of £½ m. in Filtronic. This recognition satisfied the banks, and helped in securing development funds from the Department of Trade & Industry.
It became evident in 1982 that most of the projects had a far larger potential in the USA (primarily on defence electronics), with the ratio of the USA to UK markets some 10:1. It seemed that the best way of entering the US market was to purchase an American company. Additional venture capital funds were raised to purchase a company in San Diego, California.

In order to meet security requirements a new US company was founded in 1983, using further venture capital raised in the USA: then the US subsidiary was sold to the new company. These arrangements involved many problems not previously experienced including, in one period, the employment of six different firms of lawyers.

For a time administrative responsibilities occupied 90% of Rhodes's 'waking hours' as Chairman, but he now has executive managers in place, having recruited from larger electronics companies a Managing Director and a Marketing Director. He now has time to resume technical supervision of developments.

An excellent team of innovative engineers has been built up, many of whom are ex-students of his Department at Leeds University. Some 20 or so of the senior employees are shareholders, and will benefit when the company eventually goes to the market. The Chairman also serves as Technical Director, and there is a Financial Director. A further £0·75 m. venture capital was obtained in 1985 to fund the substantial growth that the company is experiencing.

6.3 Case history: The Microsystems Group plc (*Poole, Dorset*)

In 1975, two members of the IEE, Dr Roger Harding and Dr Michael Jackson, left the Plessey Group with the intention of providing a specialist service in the field of microprocessor applications, which were then at an early stage of development. Many potential users lacked the in-house skills to adapt their electro mechanical products to the new technology.

They established Microsystems Design Ltd at Poole, and built up specialised development expertise in the fields of electronic cash registers, ticket machines, bank-note dispensers and other applications for incorporation in the products of their clients.

In 1979 larger premises were taken on a factory estate, where there was scope for further expansion. A start was made in designing and manufacturing their own products for commercial and industrial applications. This activity has expanded rapidly, and is now the greater part of the business. The name Wayfarer Ticketing Systems was adopted in 1983 for marketing a range of these products.

In 1984 a holding company was established for what were now five constituents of the business; including MDL, Wayfarer and a components distribution business that had been built up from 1979 onwards. A year later the Group acquired

Callog, the organisation that had been marketing telephone logging management systems, which MDL had manufactured.

The principal products are now ticket machines for one-man buses, adopted by 40 UK bus operators in UK, the widely used Sheriff electronic taximeter, a new staff clock system, and the Callog.

Contract services of design and manufacture are provided for various customers, which reinforces the R&D for in-house products. Modern manufacturing facilities are operated at Poole, with emphasis on fulfilling the quality-management system of BS 5750: high reliability, and an excellent back-up service to customers. A minority shareholding has been taken in a US company which markets cash-register products developed by MDL.

In 1984 Dr. Christopher Wilson became non-executive Chairman, bringing in his long experience in the computer industry, initially with Ferranti and then as MD of ICL. Harding and Jackson are joint Managing Directors, the former's particular responsibilities being sales, marketing and financial matters, and the latter looks after product development and manufacturing. They have built a strong executive team, and made available to them shareholding options. Junior technical staff have been attracted by the atmosphere of the company, and include those whose first acquaintance was through working briefly during their vacations.

The Group turnover rose steadily from just over £1 m. in 1981 to £7½ m. in 1985, with a steadily increasing proportion of profit, despite quite large fluctuations in the contributions of the several parts of the business. It is the intention to diversify further by developing new products, and by suitable acquisitions.

The Group has had the support of Barclays Bank, with Hambros Bank as its financial advisors for a placing on the Unlisted Securities Market in January 1986. The placing price valued the company at £14·3 m. The gross profit (1987) now being £4·12 m. on a turnover of £19.9 m.

6.4 Case history: Barkway Electronics Ltd. (*Melbourn, Herts.*)

The company employs some 40 staff in the design, manufacture and installation of specialised audio intercommunication systems, using hard wiring, radio and fibre-optics links. Particular applications are for security, and where there is a hazardous environment. Its recent policy has been to develop its capability to undertake major development and manufacturing projects for clients, and to update and widen its range of products, introducing microprocessor signal switching.

In particular, there has been the need to formalise procedures to satisfy the requirements of customers and agencies such as the oil producers, British Telecom, BASEEFA and NATO, and BSI's Quality System, BS 5750.

Two of the Directors are executive, Geoff Chapman (Chairman and Managing Director), and Brian Perryman (Commercial Director). The company is owned by them and a third Director, together with a Merchant Bank. The firm was recently congratulated by the Minister of State for Defence Procurement (at the time,

Mr. Geoffrey Pattie), who said:

> It is particularly your sort of company that we find exciting and challenging. you have the ability to respond to requirements and to innovate

Its present location is at the Melbourn Science Park, between Royston and Cambridge, where a property previously occupied by Cambridge Instruments Ltd. has been subdivided to serve a number of medium-sized and small tenants. Barkway have a lease on about 8000 ft^2, comprising offices, design and development, and assembly and test. All mechanical piece-part manufacture is subcontracted, as many excellent services are available within a radius of 10–30 miles.

The company has a long history, having been set up in 1967 by an engineer previously with a Norwegian manufacturer of intercom equipment. The original backer sold out in 1972 to a private investment group, after a period of losses and dependence on sub-contract wiring. Geoff Chapman joined from Pye Ltd. as marketing manager, and a situation of increasing profits was achieved.

In 1975 Chapman and Perryman attempted to obtain funds in order to buy control of the company, but without success. They were advised that at that time the company had not an adequate track record, but in the process of the attempt they learned a lot. In 1977 they were successful, with the help of another company. In 1980 they negotiated to buy out that company's interest, after a period in which it had been difficult to reconcile views on the objectives for Barkway: they wished to increase the level of R&D.

The buy out of the other company's interest was done with the help of ICFC, which took a 30% stake for £190 000. ICFC recognised that Barkway previously had problems, but had noticed the changes in its fortunes when these two people came in.

The story at that point was told in a half page of the *Financial Times.** The factory had been located at Barkway near Royston, and now needed extra space, and the move to the Science Park was made in 1982.

Further operating funds were needed, and the City accountants Robson Rhodes assisted in the preparation of a business plan. The merchant bank Warburg's took a 33% share in the equity during 1984, and a third Director provided 10% as a personal investment under the Business Expansion Scheme. By now ICFC'S share had been diluted to 10%, and Chapman and Perryman jointly held just under 50%.

To assist with financing work in progress on major contracts, Barclays Bank provided overdraft facilities, but only against the security of houses of the two executive directors. 1986 proved to be a growth period, with some substantial orders, one being for £0·25 m. Although the operating profits are excellent, there have been cash-flow problems in financing this growth. This is due to the increased work in progress, and the fact that some large and well known companies are very slow to pay.

As growth continues, Barkway will have to look for further funding, either through the Unlisted Securities Market or elsewhere.

* *Financial Times,* 4th August, 1981, p. 9

6.5 Case history: Millennium Professional and Technical Services
(Pin Green, Stevenage, Herts.)

This company was set up in 1977, supported by a nucleus of eight engineers who had done pioneer work on systems engineering at the Warren Point Laboratories.

The founder spent six months on the preparations, including attendance at a TOPS business management course. People moved over gradually, there being two staff by the end of the first year, working from home. A third joined in January 1978, and then a fourth came in as a salesman.

By 1980 there was the full initial team of eight, some doing agency work for clients and others handling projects in the company's name. The offices that had been acquired became overcrowded, and a building was erected to their own design in 1981, at the present address. By 1985 the staff was nearly 30 in number, and an extra industrial unit nearby was taken on a short lease: this provides workshop facilities for the assembly and testing of prototype systems. Full manufacture is usually done by Millennium's clients, and marketed under their names.

All work is carried out under full project control, much of it now on a turnkey basis. A typical contract would be a bespoke automatic test equipment (ATE). Millennium will design the system and write software, and buy in the 'boxes' of hardware such as oscilloscopes and data-processing equipment. These are then assembled in standard enclosures, such as the Schroff range, and then installed and commissioned on the client's site.

The company has entered the field of computer-aided design (CAD) by putting into production a 3D modelling system, originated by Leeds University. It is also in the process of undertaking its own PCB layout design instead of using a sub-contractor, and may tackle gate array design.

As an approved consultancy under DTI's MAPCON scheme, the company can undertake feasibility studies funded by a grant, and then provide the complete service of development and production.

A number of software and hardware products have been developed, to be marketed under the Millennium name. For the future, the options remain open for the balance between proprietary products and the original business of consultancy and design services for clients. There are some 50 well known names on the client list: about two-thirds of the work is for commercial end use, and one-third for Government departments.

The preferences for the future are:

- concentrate on the larger jobs (small ones are difficult to do economically)
- move from mainly software to a balance of hardware sales
- undertake some marketing and development abroad
- regularise the company structure

The initial concept of co-operative ownership was that the original team of eight would be equal partners. Now five of these remain, but occupy various functions at different levels.

The Managing Director, Dave Pearce (early 1986), said that experience has shown that it is necessary to have a more conventional executive management structure, seen to be quite separate from the functions of shareholding. Staff other than the founders have been recruited under normal conditions of employment. New recruits are mainly experienced people previously with software and systems agencies, aged about 30, plus a few direct from college. Three of the staff carry out the prototype work, there is a Marketing Director with an assistant, a Design/Services Manager, and a Buyer.

The partners initially took quite low salaries out of the business, and over nine years Millennium has continued to plough back profits, financing its expansion and the construction of its own building.

There was no need for the injection of permanent external capital, so it is unlikely to take the more usual USM route. However, there may be benefits in linking with a large complementary business, or becoming a unit in a large commercial Group.

From time to time substantial short-term Bank accommodation has been obtained, where a relatively large initial outlay is necessary in order to start on a contract (such as obtaining hardware items). Negative cash flow develops until progress payments begin to catch up with expenditure. The Bank is quite happy to help when the contract is with a well known firm or a Government Department, but would be less willing to help for speculative developments. If the company joins a large Group, they would take over the role of banker.

6.6 Case history: S I D Inc. (*Westmont, New Jersey, USA*)

This represents another example of opportunities in the USA: but in this case Peter Stoveld had the experience of the traditional immigrant, and at first he lived within sight of the Statue of Liberty, commuting daily from Staten Island to a computer-systems development task in Brooklyn.

He speaks highly of the value of what he learned in his initial work in the UK with EMI Electronics on ICL systems, followed by development of civil-engineering-industry applications with McAlpines. Recognising his need of more experience with IBM systems, he signed up in 1974 for an initial year's contract with one of the larger US system houses, then recruiting in UK as a means of acquiring experienced staff who were prepared to be mobile.

A Labour Union office in Brooklyn might be classed as 'the pits', but he decided to stay on after the first year, imported an English wife, and moved on to Ingalls' shipyard near New Orleans (he had been a keen amateur brass musician), where he tackled a production analysis job. Then the systems house moved him to an in-depth job with the Western Regional Operations Centre of American Express in Phoenix, Arizona.

In mid-1977, the couple packed their possessions into their 4-wheel drive truck, and drove back across the continent to the next posting, in Philadelphia, serving FMC's Industrial Chemical Group. He distinguished himself that winter when, after an exceptional snowfall, his '4-wheel drive' got him into the office to keep the system up and running.

FMC has requirements peculiar to the North American continent: there are some 15 plant locations and 3000 customers, and a relatively high percentage of the cost of the chemical products is represented by the freight charges. There are alternative routes across the continent, and a choice of rail/water/road carriers.

Delays of weeks were customary in updating the printed 'freight rate manuals', and an analysis showed that a high proportion of the invoiced amounts were in error, usually to FMC's disadvantage, as rates tend to rise rather than fall.

The task at FMC was to devise and implement a total order processing system, of which the freight aspect was a subsystem: it was on-line and with the database updated continuously, rather than by batch in arrears.*

Shortly after this time, interstate and infrastate trucking freight rates were being deregulated by the Government, creating a much wider range of base data on alternatives available to shippers. The system as developed has the acronym TRRIMS (transportation rate & route information management system). In order to satisfy the variety of FMC's business, from air-freighting mixed small orders of high-value items to railroading train loads of soda ash, the system had to have the capability to be personalised to suit different applications.

Other large Corporations became aware of FMC's new system, and made inquiries as to its commercial availability. By this time, Stoveld and an American colleague had gone independent and were providing professional services: they negotiated successfully for the rights to market TRRIMS to other users. The licensing agreement provided that FMC's original investment in the system would be recovered from royalties on each application, and the number of these now exceeds ten, mainly with leading chemical and oil corporations.

The new systems venture was named SID Inc. (for 'Systems and Information Dynamics'). Its initial address in 1979 was a suburban Post Office box number, and the office was a spare bedroom of Stoveld's house. In 1981 there was an opportunity to acquire an office in a professional district, vacated by an architect.

About this time Stoveld became President, having bought out his partner, whose preference was for a salaried job and regular hours. The nature of the business began to change from the provision of professional services to the development of proprietary software for further applications, and an IBM-compatible computer was installed so that this work could be done in-house.

The number of staff rose from 15 to a peak of 22 in 1982, and then was allowed to decline by attrition in a year that only broke even, with some of the staff deployed on the development and marketing of a new proprietary system of rail-fleet management. This was an investment for the future: the company had reached this point without any injection of internal capital, and it had now bought its own office property.

Overheads were kept at a minimum, and the employment of administrative

* *Distribution,* 1981, **80**, pp. 66–67 *Computer Systems Report,* Aug. 1982, **1(8)**,

staff was avoided by using the services of local accountant and law firms for payroll and similar services.

It was recognised that, for the future, it was necessary to have both a wider independent client base, and a progression of system enhancements and new products. The most recent of these is 'X-T' an international transportation management system. The company has established its identity in this particular market niche for transportation and logistics systems, and now aims for stability by diversifying its client base, and by providing a wider range of professional services and technical skills. The staff has risen to 30 (1986), and they benefit from a profit-sharing scheme that was introduced in 1981.

Peter Stoveld's impressions ten years earlier were that the level of systems work in USA was very similar to that in the UK, but the quality standards were lower, resulting in a higher ratio of debugging/writing time. He adopts a modular approach to system design, in the interests of flexibility, when customising, and he has insisted on targets of 'right first time'; he has had some successes in troubleshooting difficult existing installations, albeit by the input of extended working hours.

After hiring a variety of 'experienced' staff, he now follows a long-term policy of modest growth rate, by taking in young people at 'entry level', drawing on his personal early experience in England. SID Inc. actively co-operates in this development with the Computer Learning Centre of Airco-BOC in Philadelphia, which has diversified from its long-standing series of craft training courses in welding technology. Extensive use is made of video cassettes in training new recruits in a six-month programme, followed by a year or so of participation in project teams, to provide practical experience.

Future prospects for the business are assessed periodically – the rather specialised product range and services are probably of interest only to the *Fortune* 'top 300', but the international opportunities are exciting.

6.7 Physical facilities

The facilities available to new businesses vary from district to district, and it is wise to survey at least three options in the preferred areas, both for accommodation and services, and availability of the type of staff to be recruited if the business expands.

Professional staff are usually sensitive to the general environment, and it has been found easier to attract people to the ambience of a modern Science Park, than to an older industrial area, where, traditionally, a small business might have its start in a railway arch, or a disused shop front.

As an example of a potential location, Stevenage is representative – it is 'north of Watford' but in the home counties – it is a mature New Town, one of the first eight built in the 1950s in the south. The town has recognised the need to make extra sites available for industrial development, both in large and small units, and the Chief Executive of the Borough Council encourages inquiries.

The town's Employers' Group set up 'Stevenage Initiative' to advise people who wanted to start small businesses, and it has had the assistance in financial matters of an adviser seconded by Barclays Bank. The Director is on secondment from ICI, and three of the staff from British Aerospace, both major local employers.

A Business & Technology Centre was established in 1983, in a modern building vacated by the Printer Division of Control Data Corporation, who have remained the landlords. They had previous experience of setting up a similar Centre in the USA. The accommodation units range from 100 to several thousand square-feet of offices and light workshops, plus common services. Small secretarial, reprographic and catering enterprises provide these services for the other tenants.

By 1986 the Centre was occupied by 86 businesses, and, since the opening, 29 companies have outgrown their accommodation and moved out.* Since the beginning, assistance has been given to 137 start-ups, both low and high technology, The failure rate, at about 10%, is well below the national average.

The Borough Council issues a Register of all premises available in the Town, with names of the representative agents. Typical accommodation costs are:

- New industrial units of about 10 000 ft^2: £2–3 per ft^2 per annum; long lease with 5-year reviews
- Small office/workshop units from 1000 ft^2: £3–4 per ft^2 per annum; 15 year lease, 3-year reviews
- Units of Business & Technology Centre* from 100 ft^2: flexible leases, from 30 days
- Office accommodation adjacent to Town Centre, 500–1500 ft^2: £6–7 per ft^2 per annum

Costs in other parts of the UK may be substantially lower, but may also be higher in the Thames Valley, and in the belt of Greater London within the M25 girdle.

*The Stevenage Centre provides accommodation on the basis of a 'license fee' which covers rent, rates, electricity and heating; there is also reception and access to other services: An office of 200 ft^2 would be about £190 for a month, and a 1000 ft^2 unit at the annual rate of £7/ft^2 (plus VAT).

Chapter 7

References and Bibliography

Introduction

The Reading Lists, at the end of each main Chapter of this book, are a short selection from the extensive management literature and include suitable inexpensive 'paperbacks'.

References in the text make use of the 'Harvard system' thus: Drucker (1954)

They will be found under the appropriate sub-sections on the following pages: the alphabetical listing also provides an author index.

When a particular subject is pursued in greater depth, access to the books or articles can be obtained with the assistance of the technical and commercial Libraries, and the notes in Chapter 5 (5.3.5) on Desk Research will be useful. The staff of the Libraries of the IEE and other principal Engineering Institutions are extremely helpful, and members of these bodies and the BIM can borrow books by post.

Photocopies of articles can be ordered either by post or telephone. Librarians prefer the latter personal contact as it enables them to resolve queries at once. They will despatch urgent requests to members and notify the cost of copying. Both the IEE and BIM have facilities for facsimile transmission. Charges are of the order of £1 handling charge per item, 10p./page for copies, and £1/page for facsimile transmission.

If the reader's trail leads more deeply into an aspect of management, a helpful source of guidance is the series of Management Abstracts, issued by ANBAR in association with BIM.

Abbreviations:

IEE Proc. : IEE Proceedings
IEEJ. : IEE Journal
LRP : Long Range Planning
J.ORS : Journal of Operational Research Society
SLRP : Society of Long Range Planning (recently SPS: Strategic Planning Society)

7.1 Organisation

ACARD (1982): 'Facing international competition: Report' (London, HMSO)
ACKOFF, R. L., GHARAJEDAGHI, J., and FINNEL, E. V. (1984): 'A guide to controlling your corporation's future', (New York, Wiley) 165 pp.
ANSOFF, H. I. (1968): 'Corporate strategy' (Harmondsworth, Penguin) 205 pp.
ANSOFF, H. I. (1979): 'Strategic management' (London, Macmillan)
ANSOFF, H. I., BOSMAN, A., and STORM, P. M. (Eds.) (1982): 'Understanding and managing strategic change' (Oxford, North Holland) 251 pp.
ARGENTI, K. (1984): Argenti Systems Ltd, Norwich
AULETTA, K. (1984): 'The art of corporate success – the story of Schlumberger' (New York, Putnam) 184 pp.
INST. PROD. ENGRS. (1957): 'Automatic production – Change and control' (Automation Conf., Harrogate) (London, Spon), 228 pp.
BAGRIT, LORD (1965): 'The age of automation ' (Reith Lecture, 1974) (Harmondsworth, Pelican) 92 pp.
BAKER, J. W. (1985): 'When the engineering has to stop' *IEE Proc.* **132** Pt. A, pp. 193–198
BANKS, J. (1983): 'The engineer in the middle ground of manufacture: *IEE Proc.* **130**, Pt. A. pp. 7–18
BARNARD, C. I. (1968, 1938): 'The functions of the executive' (Cambridge, Mass., Harvard Univ Pres (reprinted)) 334 pp.
BARNETT, CORRELLI (1986): 'The audit of war – the illusion and reality of Britain as a great nation' (London, MacMillan) 359 pp.
BEGGS, J. M. (1984): 'Leadership – the NASA approach', *LRP*, **17**, pp. 12–24, April
BELBIN, R. M. (1982): 'Management teams – why they succeed or fail' (Aldershot, Gower) 157 pp.
BERGEN, S. A. (1984): 'Catastrophe model of the engineering design process' *IEE Proc.* **131**, pp. 181–184
BESTWICK, P. F., and LOCKYER, K. (1982): 'Quantitative production management' (London, Pitman) 442 pp.
BIM (1985): 'Improving management performance' (London, BIM) 42 pp.
BIM Reading List: 'Automated design and production'
BIM Checklist no 23: 'Mergers and takeovers'
BIM Checklist no 33: 'Physical distribution'
BIM Checklist no 85: 'Organisation structure and organisation theory' (Corby, BIM Management Information Centre)
BLACK, H. S. (1934): 'Stabilised feedback amplifiers' *Bell Syst. Tech. J.* **13**, January, pp 1–18
BLUMLEIN, A. D., BROWNE, C. O., DAVIS, N. E., and GREEN, E. (1938): 'The Marconi-EMI television system' *J. IEE,* **83**, pp. 758 *et seq.*
BOWERS, B. (1985): 'The first century of the Wiring Regulations' *IEE Proc.,* **132** pt. A pp. 498–502
BRAGG, J. P. (1982): 'Priority base budgeting: a practcal zero-base approach to manufacturing overheads: *IEE Proc.,* **129**, Pt. A pp. 76–80
BRAY, D. W., CAMPBELL, R. J., and GRANT, D. L. (1974): 'Formative years in business – a long-term A&TT study of managerial lives: (New York, Wiley) 236 pp.
BRYSON, SIR LINDSAY, (1982): 'Large-scale project management' *IEE Proc.* **129** Pt. A, pp. 625–633
BURBRIDGE, R. N. G. (1984): 'Some art, some science and a lot of feedback', *IEE Proc.,* **131** Pt. A pp. 24–37
BUNN, D. W., and SEIGAL, J. P. (1983): 'Forecasting the effects of television programming upon electricity loads' *J.ORS,* **34**, pp. 17–25

BURNS, T. and STALKER, G. M. (1961): 'The management of innovation' (London, Tavistock)

BURNS, R. W. (1986): 'Seeing by electricity' *IEE Proc.* **133**, Pt. A, pp. 27–37

Business Schools (1982): *IEE Proc.*, **129**, Pt. A (Special Issue) pp. 205–284

CARSBERG, B. (1986): 'OFTEL: a new approach to regulation', *IEE Proc.* **133**, Pt. A, pp. 152–158

CHANDLER, A. (1986): 'A system engineering approach to software development', *IEE Proc.* **133**, Pt. A, pp. 355–357

CHANDLER, A. D., and SALSBURY, S. (1971): 'Pierre S DuPont and the making of the modern Corporation' (New York: Harper & Row) 720 pp.

CLELAND, D. I., (Ed.) (1984): 'Matrix management systems handbook' (New York, Van Nostrand Reinhold) 757 pp.

CLELAND, D. I. (Ed.) (1983): 'Project management handbook' (New York, Van Nostrand Reinhold) 725 pp.

CLELAND, D. I., and KING, W. R. (Eds.) (1983): 'Systems analysis and project management' (New York, McGraw Hill) 490 pp.

COALES, J. F. (1983): 'The role of the engineer in a free economy', *IEE Proc.* **130** Pt. A, pp. 393–400

Colloquium (1986): 'Systems Engineering' *IEE Proc.* **133** Pt. A. pp. 329–360

Colloquium (1986): 'Project management lessons learned from the Drax completion project' in 'Project Management' (London, IEE/Peter Peregrinus) – (1988) (editor R. N. G. Burbridge)

CONSTABLE, J. (1986): 'Diversification as a factor in UK industrial strategy', *LRP*, **19**, pp. 52–60, February

CONWAY, H. G. (1951): 'Industrial design and its relation to machine design', *Proc. I.Mech. E.*, **164**, pp. 177–194

COOLEE, T. E. (1986): 'Mergers and acquisitions' (Oxford, Blackwell)

COPEMAN, G., MOORE, P., and ARROWSMITH, C. (1984): 'Shared ownership: how to use capital incentives to sustain business growth' (Aldershot, Gower) 256 pp.

CRITCHLEY, T., SNOOKS, B.K., ROBERTSON, A.G., MURPHY, J.H., GRIFFITHS, F. and BRYSON, Sir Lindsay: 'Professional Purchasing – a Key Skill for Engineers' in: (1988) 'Engineering Manufacturing Systems' (London: Peter Peregrinus)

DEVINE, W. W. (1983): 'From shafts to wires: historical perspectives on electrification', *J. Econ. History*, **43**, pp. 347–372

DRUCKER, P. F. (1954): 'The practice of management (London, Heinemann/Pan) 480 pp.

DRUCKER, P. F. (1971): 'The age of discontinuity: guidelines to our changing society' (London, Heinemann/Pan) 477 pp.

DRUCKER, P. F. (1974): 'Management: tasks, responsibilities, practices' (London, Heinemann) 840 pp.

DRUCKER, P. F. (1978): 'Adventures of a bystander' (New York, Harper & Row) 344 pp.

DRUCKER, P. F. (1985): 'Innovation and entrepreneurship' (London, Heinemann)

DSIR (1956): 'Automation' (London, HMSO) 106 pp. (comprehensive bibliography)

EIU (1982): 'Management buy-outs' Report No 115 (London, Economic Intelligence Unit)

FAROWICH, S.A. (1986): 'Communicating in the technical office', *IEEE Spectrum*, **23**, pp. 63–67

FAYOL, H. (1929): 'Administration Industrielle et General – prevoyance, organisation, commandement, co-ordination, control' (Geneva, International Management Institute) (English translation, first published in French in 1916)

FONTAINE, L., WALKER, J. W., and SPENCER, W. R. (1958): 'Management control of small engineering firms: I Bulding and controlling a management team; II Control of production engineering; III Control of finance and costs', *J. I.Mech.E.*, **172**, pp. 361–416

FRAME, Sir A. (1986): 'A client's view of project management', *Financial Times*, 4 Sept. 1986

FREEMAN, C. (1985): 'The economics of innovation', *IEE Proc.,* **132** Pt. A, pp. 213–221

FULTON, Lord (1968): 'The Civil Service – Report of the Committee', Cmnd. 3638, vol 1, and vols 2–5 (London, HMSO) Vol. 1, paras. 153–156

GAITHER, N. (1984): 'Production and operations management – a problem solving approach', (New York, Dryden, 2nd edn.) 768 pp.

GENEEN, H. S. (1985): 'Managing' (London, Granada) 225 pp.

GEORGE, W. W. (1983): 'Contrasts in management styles – Europe and America', *IEE Proc,* **130** Pt. A., pp. 288–291

GILBERT, K. R. (1965): 'The Portsmouth blockmaking machines' (London, HMSO/Science Museum) 42 pp.

GILLIAN, P. (1985): 'The business of quality' *Brit Telecom J,* Summer 1985, pp. 2–4

GIRLING, D. S. (1980): 'Changes in quality management of electronic components', *IEE Proc.,* **127** Pt. A, pp. 407–414

GREEN, A. (1983): 'The management of a CAE project', *Electronics & Power,* **29**, pp. 78–80

HADNAM, R. G. A. (1985): 'Strategic planning in purchasing', *Purchasing and Supply Management (UK),* November, pp. 35–41

HARVARD LIBRARY: 'Management of projects and programmes' (15 collected articles from Harvard Business Review) (London, Heinemann) ref. 18.011

HAWTHORNE, E. P. (1970): 'MbO in research and development', See Humble (1970), pp. 102–124

HENDRY, M. (1986): 'Applying strategic planning at the national level', *Electronics & Power,* **32**, pp. 353–4

HOLROYD, P. (1983): 'Applications of network theory in social systems', *IEE Proc.,* **130** Pt. A pp. 273–280

HORNER, R. W. (1987): 'The Thames Barrier', *IEE Proc.,* 134, Pt. A pp 752–760.

HUDSON, P. G., WEBB, S. and CHANDLER, J. R. (1986): 'End user project management for Holset FMS', *Computer-Aided Engg. J.,* **3**, pp. 144–149

HUMBLE, J. W. (1965): 'Improving management performance' (London, BIM) 63 pp.

HUMBLE, J. W. (Ed.) (1970): 'Management by objectives in active' (London, McGraw-Hill) 294 pp.

HUMBLE, J. W. editor (1973): 'Improving the performance of the experienced manager' (London, McGraw Hill) 346 pp.

HUSSEY, D. E. (1978): 'Corporate Planners' yearbook' (Oxford, Pergamon) 243 pp.

HUSSEY, D. E. (1982) 'Corporate planning: theory and practice' (Oxford, Pergamon) 523 pp.

HUSSEY, M. (1987): 'Quantitative methods for managers: an engineering perspective' (London, IEE/PPL) (to be published)

I. PROD. E. (1955): 'The Automatic Factory' Conference report (London, Spon) 226 pp.

I. PROD. E. (1957): 'Automatic production – change and control' Conference report (London, I. Prod. E.) 252 pp.

JAY, A. (1971): 'Corporation man' (New York, Random House) 304 pp.

JOHNSON, C. (1982): 'MITI and the Japanese Miracle' (Stanford: Stanford Univ. Press)

JONES, H. (1983): 'Preparing company plans: a workbook for effective corporate planning' (Aldershot, Gower, 2nd edn.) 299 pp.

JURAN, J. M. (1935): 'Inspectors' errors in Quality Control', *Mechanical Engineering (US),* October, pp. 643–644

KAMINSKI, M. A. (1986): 'Protocols for communications in the factory' *IEEE Spectrum,* **23**, pp. 56–62, April

KEEBLE, S. P. (1984): 'University education and business management from the 1890's to the 1950's – a reluctant relationship'. Doctoral thesis London School of Economics: Business History Unit

KELLEY, A. J. (Ed.) (1982): 'New dimensions in project management' (Lexington,Mass., D. C. Heath) 210 pp.

KERZNER, H. (1981): 'Project management for executives' (New York, Van Nostrand Reinhold) 716 pp.

KHARBANDA, D. P., and STALLWORTHY, E. A. (1982): 'How to learn from project disasters' (Aldershot, Gower) 274 pp.

KIDDER, T. (1981): 'The soul of a new machine' (Harmondsworth, Allen Lane/Penguin) 256 pp.

KNIGHT, K. (Ed.) (1982): 'Matrix management: a cross-functional approach to organisation' (Aldershot, Gower, 2nd edn.)

LIVINGSTONE, A. W. (1984): 'An inside view of Japan Inc.', *IEE Proc.* **131** Pt. A, pp. 335–340

LANSDOWN, J., and MAVER, T. (1984): 'CAD in architecture and buildings' *Computer-Aided Design,* **16**, pp. 146–160

LAWLESS, J. (1986): 'The Weinstock way', *Business,* Ap., pp. 77–82

LELE, M. M., and KARMARKER, U. S. (1983): 'Good product support is smart marketing', *Harvard Business Review,* Nov.–Dec., pp. 124–132

LESTER, A. (1982): 'Project planning and control' (London, Butterworth) 183 pp.

LILLEY, S. (1957): 'Automation and social progress' (London, Lawrence & Wishart) 224 pp.

LLEWELLYN, R. V. (1985): 'CAE – a user's case history and reflections', *Computer-Aided Engg. J.,* **2**, Oct. pp. 144–149

LOCK, D. (1977): 'Project management' (Aldershot, Gower) 260 pp.

LOMAS, T. (1983): 'Managing for quality', *British Telecom J.,* Winter 1983/84, p. 33

LORENZ, C. (1982): 'Roots of the British malaise' *Financial Times,* Sept.

LORENZ, C. (1986): 'The two cultures', *Financial Times* (Management page) 17 January

McCAFFERTY, D. N. (1980): 'The National Freight buy-out' (London, Macmillan) 208 pp.

McPHATER, N. S. (1985): 'Bridging the culture gap between engineers and software developers', *Computer-Aided Engg. J.* **2**, pp. 84–88

M'PHERSON, P. K. *et al.* (1986): 'System engineering: its nature and scope' *IEE Proc.* **133**, Pt. A, pp. 329–358

MALPAS, R. (1982): quoted by Knight (1982)

MALPAS, R. (1985): 'The technical review – engineering statements for investors' *IEE Proc.* **132**, Pt. A, pp. 481–4

MARCUS, C. L. (1979: 'Codes for customer relations' *PO Telecom J.,* **31**, (2) Summer, pp. 26–27

MAYER, T. (1973): 'Management for the benefit of people and enterprise', *IEE Proc.,* **120**, pp. 146–150

MILLS, C. A. (1983): 'Managing for quality' *IEE Proc.,* **130** Pt. A, pp. 58–62

MOORE, G. (1986): 'Manufacturing automation protocol (MAP) – mapping the factory of the future', *Electronics & Power,* **32**, April, pp. 269–272

MORLEY, W.B. (1986): 'The privatization of British Telecom – its impact on management' *LRP,* **19**, December, pp. 124–129

NAISBITT, J., and ABURDENE, P. (1985): 'Reinventing the corporation – transforming your job, and your company for the new information society' (London, MacDonald) 308 pp.

NEW, C. (1986): *Sunday Times,* 15 June, p. 73

NEDO: (1984): 'Computers in production control' (London, HMSO) 35 pp.

NORMANN, R. (1984): 'Service management: (Corby, BIM)

NYSTROM, P. C., and STARBUCK, W. H. (Eds.) (1981): 'Handbook of organisational design'. Vols. 1 and 2 (London, Oxford University Press) 588 pp. and 540 pp.

O'HARA, S. (1985): 'How Motorola got to be number one', *New Management (USA),* Autumn, pp. 7–14

ODIORNE, G. S. (1979): 'Mbo II – a system of management leadership' (Belmont, Calif., Fearon Pitman) 360 pp.

OULD, J. (1986): 'Controlling subsidiary companies' (Cambridge, Woodhead-Faulkner) 240 pp.

PASCOE, B. J. (1972): 'The Royal Naval Supply and Transport Service' see HUMBLE, J. W. (Ed.) (1970), pp. 61–77

PEARSON, A. W. (1983): 'Project management in R & D', *Proc I.Mech.E.*, **197**, November

PERRIN, R. (1985): *SLRP Newsletter*, November, p. 9.

PETERS, R. J. and WATERMAN, R. H. (1982): 'In search of excellence: lessons from America's best-run companies' (New York, Harper & Row) 360 pp., (and in paperback)

PINCHOT, G. (1985): 'Intrapreneuring – why you don't have to leave the Corporation to become an entrepeneur' (New York, Harper & Row) 236 pp.

PUGH, D. S., HICKSON, D. J., and HININGS, C. R., (Eds.) (1984): 'Organisation theory – selected readings' (Harmondsworth, Penguin, 2nd ed.) 447 pp.

PYHRR, P. A. (1973): 'Zero-base budgeting: a practical management tool for evaluating expenses' (New York, Wiley)

RACAL (1980): Quoted in Special Issue on Management, *IEE Proc.*, **127** Pt. A p. 582

RADAR, (1985): Special Issue on Historical Radar, *IEE Proc.*, **132**, Pt. A pp. 325–444

REDDIN, W. J. (1971): 'Effective MbO' (London, HIM) 290 pp.

REINHARDT, W. A. (1984): 'An early warning system for strategic planning', *LRP*, **17**, Oct. pp. 25–34

ROGERS, T. G. (1980): 'Plessey's evolving management philosophy', *IEE Proc.*, **127**, Pt. A, pp. 600–603

RUBENSTEIN, A. H. (1985): 'Trends in technology management' *IEEE Trans.* **EM-32**, pp. 141–143

SARGROVE, J. A. (1947): 'New methods in radio production' *J. Brit. IRE*, **3**, p. 1

SCHUMACHER, E. F. (1973): 'Small is beautiful – economics as if people mattered' (New York, Harper & Row) 290 pp.

SCHROEDER, R. C. (1982): 'Operations management – decision making' (New York, McGraw Hill) 681 pp.

SEYNA, E.J. (1986): 'MbO: the fad that changed management', *LRP*, **19**, December, pp. 116–123

SHEANE, D. (1977): 'The company-wide matrix': chap 3 in KNIGHT, K. (Ed.) 'Matrix Management' (Aldershot, Gower)

SHELDON, O. (1923): 'The philosophy of management: (London, Pitman)

SHORTLAND, M. G. and PRINCE, G. (1985): 'Advanced manufacturing technology' (London, Jordan)

SLOAN, A. P. (1963): 'My years at General Motors' (London, Pan) 470 pp.

SMITH, SIR A. (1986): 'Engineering and industry – the hidden hunger', *IEE Proc.*, **133** Pt. A pp. 630–636

SPICKERNELL, D. G. (1983): 'Design for quality' *IEE Proc.* **130** Pt. AS pp. 159–161

STUCKENBRUCK, L. C. (1981): 'Implementation of project management: the professional's handbook' (New York, Addison-Wesley)

TAKEI, F. (1986): 'Engineering quality improvement through TQC activity', *IEE Trans.*, **EM 33**, pp. 92–95

TAYLOR, D. E. (1979): 'Strategic planning as an organisational change process', *LRP*, **12**, October, pp. 45–53

TAYLOR, B. and HUSSEY, D. (1982): 'The realities of planning' (Oxford, Pergamon) 252 pp.

TAYLOR, B. and REDWOOD, H. (1982): 'British Planning Databook', *LRP*, **15**, pp. 1–160: October, also as book (Oxford, Pergamon, 1983) 168 pp.

TAYLOR, F. W. (1910): 'Shop management' (paper to ASME, 1903) (New York, Harper Bros)

THOMPSON, SIR P. (1985): 'The NFC buy-out: a new form of industrial enterprise', *LRP*, October, **18**, pp. 19–27

TISDALL, P. (1982): 'Agents of change: (London, Heinemann) 163 pp. (p.32)

TOMLINSON, H. (1982): 'Measures for success', *British Telecom J*, Winter 1982/83 pp. 12–15

TOMLINSON, H. (1983): 'Productive labour', *IEE Proc.* **130** Pt. A pp. 19–30
TUGENDHAT, C. (1971): 'The multinationals' (London, Eyre & Spottiswoode) 243 pp.
TURNER, G. (1985): 'GEC's Pragmatic planners', *LRP,* **18,** February, pp. 12–18
TWISS, B. (1987): 'Business for engineers' (London, IEE/Peter Peregrinus) (to be published)
TWISS, B. and JONES, H. (1978): 'Forecasting technology for planning decisions' (London, Macmillan) 320 pp.
TWISS, B. (1988): 'Business for engineers' (London, IEE/Peter Peregrinus) 209 pp.
TWISS, B. (1982): 'Socioeconomic significance of microelectronics', *IEE Proc.,* **129** Pt. A, pp. 237–243
URWICK, L. (1956): 'The golden book of management – 70 pioneers' (London, Newman Neame)
WEARNE, S. H. (1970): 'Product and policy responsibilities in industry' *Management Decision,* Winter, pp. 32–35
WEBB, I. (1985): 'Management buy-outs: a guide for the prospective entrepreneur' (Aldershot, Gower) 192 pp.
WELLEMIN, J. H. (1984): 'Handbook of professional service management' (Bromley, Chartwell-Bratt) 225 pp.: also to be published (1987) 'Professional service engineering'
WRIGHT, M. and COYNE, J. (1985): 'Management buy-outs' (London, Croom Helm) 256 pp.
ZENTNER, R. D. (1982): 'Scenarios, past present and future', *LRP,* **15,** June, pp. 12–20

7.2 Managing people

ACAS (1980): 'Industrial relations handbook' (London, HMSO) 354 pp.
ADAIR, J. (1983): 'Effective Leadership – a self-development manual' (Aldershot, Gower/Pan) 228 pp. (Mike Brearley's contribution, in an expanded article, in *Director,* June 1986, pp. 24–26)
AMBLER, E. (1985): 'Here Lies – an autobiography' (London, Weidenfeld/Fontana) 234 pp.
ARGYRIS, C. (1957): 'personality and organisation' (New York, Harper & Row)
ARGYRIS, C. (1970): 'Intervention theory and method' (Reading, Mass., Addison-Wesley)
ARTHUR, R. (1984): 'The engineer's guide to better communication' (Glenview, Ill., Scott, Foresman) 125 pp.
ARMSTRONG, M. (1984): 'A handbook of personnel management practice' (London, Kogan Paul, 2nd edn.) 416 pp.
AULETTA, K. (1984): 'The art of corporate success' (New York, Putnam) 184 pp.
BAGRIT, L. (1964): 'Automation' (Reith Lectures) (London, BBC Publications); also as BAGRIT, L. (1965): 'The age of automation' (Harmondsworth, Penguin) 92 pp.
BARNARD, C. (1938): 'The functions of the executive: (Cambridge, Mass.; Harvard University Press) 334 pp. (reprinted 1968)
BASS, B. (1985): 'Leadership and performance beyond expectations' (New York; Free Press)
BEGGS, J. M. (1984): 'Leadership – the NASA approach' *LRP* **17,** April pp. 12–24
BENNIS, W., and NANUS, B. (1985): 'Leaders: the strategies for taking charge' (New York, Harper & Row) 244 pp.
BITTEL, L. R. (1984): 'Leadership – the key to management success' (New York, Franklin Watts) 201 pp.
BLAKE, R. R. and MOUTON, J. S. (1964): 'The managerial grid' (Houston, Gulf Publishing)
BROWN, W., and JAQUES, E. (1965): see Jaques
BURBRIDGE, R. N. G. (1984): 'Some art, some science and a lot of feedback', *IEE Proc.,* **131** Pt. A, pp. 24–37
BUTLER, D., and CAVANGH, D. (1983): 'The British General Election, 1983' (London, Macmillan)
BURNS, T. and STALKER, G. M. (1961): 'The management of innovation' (London, Tavistock) 269 pp.
CASEY, D., and PEARCE, D. (1977): 'More than management development' action learning at GEC: (Aldershot, Gower) 160 pp. (foreword by Lord Weinstock)

CLUTTERBUCK, D. (Ed.) (1985): 'New patterns of work' (Aldershot, Gower) 160 pp.
CLUTTERBUCK, D., and HILL, R. (1981): 'The re-making of work' (London, Grant McIntyre) 216 pp.
COHEN, I. H. (1986): 'The responsibility for quality' (Colloquium 1985), *IEE Proc.* **133** Pt. A, pp. 369–374
COOPER, B. M. and BARTLETT, A. F. (1980): 'Industrial relations' (London, Heineman) 320 pp.
DAVIS, L. E., and TAYLOR, J. C. (1972): 'Design of jobs' (Harmondsworth, Penguin) 479 pp.
DAVIS, S. M. (1987): '2001 Management—managing the future now' (London: Addison-Wesley), 243 pp. (titled in USA 'Future perfect')
DRAX (1988): 'Colloquium on Drax completion project' in 'Project Management' (London, IEE/Peter Peregrinus) p. 148 (editor, R.N.G. Burbridge)
DRUCKER, P. F. (1954): 'The practice of management' (London, Heinemann/Pan) 479 pp.
DRUCKER, P. F. (1973): 'Management: tasks, responsibilities, practices' (London, Heinemann) 839 pp.
DUERR, C. (1971): 'Management kinetics – on communication' (London, McGraw Hill) 215 pp.
FLETCHER, W. (1980): 'Meetings, meetings' (London, Michael Joseph)
FORD, R. N. (1968): 'The art of reshaping jobs' and 'Motivating people' *Bell Telephone Mag.* **47**, July/Aug pp. 2–9 and Sept/Oct pp. 29–32
FORD, R. (1969): 'Motivation through the work itself' (New York, American Management Association)
FULTON (LORD) (1968): 'The Civil Service: Report of Committee' Vol 1 (London, HMSO) 206 pp.
GARNETT, J. (1983): 'On doing good – creativity and vocation in industry' (Northumberland, Stocksfield) 16 pp.
GOLDSMITH, W., and CLUTTERBUCK, D. (1984): 'The winning streak' (London, Weidenfeld & Nicolson) 204 pp.
GRAY, I. (1979): 'The engineer in transition to management' (New York, IEEE Press/John Wiley) 127 pp.
GROVE, A. S. (1983): 'High output management' (New York, Random House) 235 pp.
HANDY, C. B. (1981): 'Understanding organisations' (Harmondsworth, Penguin, 2nd edn.) 480 pp.
HANDY, C. B. (1984): 'The future of work – a guide to changing society' (Oxford, Blackwell) 201 pp.
HERZBERG, F. W., MAUSNER, B., and SNYDERMAN, B. (1957): 'The motivation of work' (New York, Wiley)
HICKMAN, C. R., and SILVA, M. A. (1984): 'Creating excellence' (London, Allen & Unwin) 305 pp.
HUMBLE, J. W. (1970): 'Management by objectives in action' (Maidenhead, McGraw Hill) 294 pp.
HUMBLE, J. W. (Ed.) (1973): 'Improving the performance of the experienced manager' (Maidenhead, McGraw Hill) 346 pp.
IEE Professional Brief (1986): 'Career Development', 16 pp.: and also see IEE NEWS, January and February 1987
IRVINE, A. S. (1971): 'Improving industrial communication – a guide' (London, Gower/Industrial Society) 310 pp.
JAQUES, E., and Brown, W. (1965): 'Glacier project papers' (London, Heinemann) 277 pp.
JAY, A. (1969): 'Management and Machiavelli' (New York, Random House)
JAY, A. (1970): 'The new oratory' (New York, American Management Association) 133 pp.
JAY, A. (1971): 'Corporate man' (New York, Random House) 305 pp.
JEFFERIES, J. R., and BATES, J. D. (1983): 'Meetings, conferences and audio-visual presentations' (Maidenhead, McGraw Hill)
JENKINS, W. D. (1974): 'A review of leadership studies, with particular reference to military problems' *Psychological Bull.*, **44**,

JOHNSTON, D. L. (1978): 'Scientists become managers – the T-shaped man' *IEEE Engineering Management Rev.*, **6**, pp. 67–8,

JOHNSTON, D. L. (1988): 'Engineering contributions to the evolution of management practice'. IEEE Engineering Management Trans. (to be published)

KIDDER, T. (1981): 'The soul of a new machine' (London, Allan Lane) 293 pp.

LAKEIN, A. (1985): 'How to get control of your time' (Aldershot, Gower) 224 pp.

LEVINSON, H. (1980): 'Executive' (Cambridge, Mass., Harvard University Press, 2nd edn.)

LIKERT, R.(1967): 'The human organisation' (New York, McGraw Hill)

LOCK, D., and FARROW, N. (1983): 'The Gower Handbook of Management' (Aldershot, Gower, 2nd edn.) 1200 pp.

LUPTON, T. (1980): 'Management and the social sciences' (Harmondsworth, Penguin) 176 pp.

McCLELLAND, D. (1961): 'The achieving society' (New York, D Van Nostrand)

McGREGOR, D. (1960: 'The human side of enterprise' (New York, McGraw Hill) 246 pp.

MARGERISON, C., and McCANN, R. (1985): 'How to lead a winning team' (Bradford, MCB) 80 pp.

MASLOW, A. H. (1954): 'Motivation and personality' (New York, Harper & Row)

NAISBILL, J. (1984): 'Megatrends – ten new directions' (London, Macdonald: New York, Warner – 1982) 290 pp.

ODIORNE, G. S. (1979): 'MBO II – a system of managerial leadership for the 80s' (Belmont, Calif.; Fearon Pitman) 360 pp.

PAUL, W. J. and ROBERTSON, K. B. (1970): 'Job enrichment and employee motivation' (London, Gower)

PAUL, W. J., and ROBERTSON, K. D. (1969): 'Learning from job enrichment' (London, ICI Central Personnel Department) 59 pp.

PEACH, L. H. (1983): 'Employee relations in IBM' *Employee Relations,* **5** (3) pp. 17–20

PETERS, T. J., and WATERMAN, R. H. (1982): 'In search of excellence: lessons from America's best-run companies' (New York, Harper & Row) 360 pp., (and in paperback)

PETERS, T. J., and AUSTIN, N. (1985): 'A passion for excellence' (London, Collins) 437 pp.

PETERS, T. J. (1987): 'Thriving on Chaos' (London, Macmillan), 561 pp.

PINCHOT, G. (1985): 'Intrapreneuring' (New York, Harper & Row) 236 pp.

PRIESTLEY, J. B. (1932): 'Dangerous Corner' (London, Samuel French, acting edition) 76 pp.

QUINN, J. J. (1986): 'Monitoring the technical environment' *IEE Proc.* **133** Pt. A, September, pp. 361–364

REDDIN, W. J. (1970: 'Mangerial effectiveness' (Maidenhead, McGraw Hill)

REDDIN, W. J. (1971): 'Effective MBO' (London, Management Publications/BIM) 289 pp.

REVANS, R. W. (1971): 'Developing effective managers' (Harlow, Longmans)

RODGERS, B. (1986): 'The IBM way' (New York, Harper & Row) 235 pp.

ROETHLISBERGER, F. J., and Dickson, W. J. (1939): 'Management and the worker' (Boston, Harvard University Press) 615 pp.

ROETHLISBERGER, F. J. (1942): 'Management and morale' (Cambridge, Mass., Harvard University Press) 194 pp.

ROSE, M. (1970): 'Industrial behaviour: theoretical development since Taylor' (Harmondsworth, Penguin) 304 pp.

SCOTT, B. (1984): 'Communication for professional engineers' (London, Thomas Telford) 240 pp.

SEYNA, E. J. (1986): 'MbO: the fad that changed management', *LRP,* **19**, December, pp. 116–123

SINGER, E. J., and RAMSDEN, J. (1972): 'Human resources – obtaining results from people at work' (Maidenhead, McGraw Hill) 197 pp.

SLOAN, A. P. (1963): 'My years with General Motors' (London, Pan Books ed.) 480 pp.

TAYLOR, R. L. (1975): 'The technological gatekeeper', *R & D Management* (*UK*), **5**, pp. 235–238

TISDALL, P. (1982): 'Agents of change: the development and practice of management consultancy' (London, Heinemann) 163 pp.

TOFFLER, A. (1970): 'Future shock: the ways we adapt to change' (London, Bodley Head/ Pan) 517 pp.

TOMLINSON, H. (1983): 'Productive labour' *IEE Proc.*, **130**, Pt. A, pp. 19–30
TOWNSEND, R. (1970): 'Up the Organisation' and (1984): 'Further up the Organisation' (London, Joseph)
VAN HAM, K. and WILLIAMS, R. (1986): 'Quest for Quality at Philips', *LRP*, **19**, December, pp. 25–30
VROOM, V. H., and DECI, E. L. (Eds.) (1980):' Management and motivation' (Harmondsworth, Penguin) 400 pp.
WALTERS, P. (Sir) (1986): 'The £41 billion strategist' *Director*, **39**, pp. 24–26
WHINCUP, M. (1980): 'Modern employment law' (London, Heinemann) 320 pp.

7.3 Managing financial resources

7.3.1 General concepts and background: Books

ASHTON, R. K. (1983): 'UK financial accounting Standards' (Cambridge, Woodhead-Faulkner) 234 pp.
BROWN, R. G., and JOHNSTON, K. S.(1963): 'Paciolo on accounting'' Quoted in Encyclopaedia Britannica (1974), 15th edition, volume V, p. 43
FRANKS, J. R., BROYLES, J. E., and CARLETON, W. T. (1985): 'Corporate finance, concepts and applications' (London, Wadsworth)
GLAUTIER, M. W. and UNDERDOWN, B. (1986): 'Accounting theory and practice' (London, Pitman), 3rd edition, 672 pp.
HARDCASTLE, A., and RENSHALL, M. (1985): 'Financial accounting under the Companies Act 1981' (London, Peat, Marwick & Mitchell) 148 pp.
HARVEY, M., and KEER, F. (1983): 'Financial accounting theory and Standards' (UK, Prentice Hall, 2nd edn.)
JOHNSON, H., and WHITTAM, A. (1984): 'A practical foundation in accounting' (London, Allen & Unwin, 2nd edn.)
de METZ, R. (1984): 'Off balance sheet finance' (Corby, BIM)
PAGE, J. (1982): 'Accounting and information systems' (UK, Prentice Hall)
PIZZEY, A. (1980): 'Accounting and finance' (New York, Holt, Reinhardt & Winston, 2nd edn.)
POCOCK, M. A., and TAYLOR, A. H. (1981): 'London, Gower) 446 pp.
SAWYER, A., and WALK, R. (1984): 'Accountancy – questions and answers '' (I of Bankers' stage 2 exam) (London, Financial Training Publications)
WOOD, F. (1984): 'Business Accounting' Vol. I and II (London, Longman, 4th edn.)
ICA (1984): 'Selected accounting standards – interpretation problems explained) (London: ICAEW) ref 657 Published for Accounting Standards Committee at Institute of Chartered Accountants
Harvard Business Review Library (series) (London, Heinemann). Reprinted selections of key articles on the several aspects of management, including: Pt. 3 – Managerial economics (quantitative approaches to decision making); Pt. 5 – Finance; Pt. 7 – Planning and control
'Financial management handbook' (London, Kluwer-Harrap) 800 pp. (loose-leaf handbook, with up-date service)

7.3.2 Accountancy for the non-accountant: Books

BATTY, J. (1976): 'Accounting for research and development' (London, Business Books)
SSAP 13 (1979): 'Accounting for research and development' (London: ICAEW, for Accounting Standards Committee)
FRENCH, D. (1980): 'Dictionary of accounting terms' (London, ICA)
GOCH, D. (1986): 'Finance and accounts for managers' (London, Kogan Page, rev. ed.)
ILO (1985): 'How to read a balance sheet'. Programmed book (Geneva/London), ILO, 2nd edn.) 213 pp.

HAYES, R. S. (1980): 'Simplified accountancy for engineering and technical consultants' (New York, Wiley)
JONES, D. M. C. (1978): 'Management accounting for non-financial managers' (London, BIM) 47 pp.
MOTT, G. (1980): 'Accounting for non-accountants' (London, Pan) ref. 282352
NORKETT, P. (1982): 'Accountancy for non-accountants'. Vol. I and II (London, Longman)
OLDCORN, R. (1984): 'Understanding company accounts' (London, Pan Breakthrough Books)
PARKER, R. H. (1980): 'Understanding company financial statements' (London, Pelican) ref. 224211
ROBINSON, K. R. (1985): 'Accounting definitions for the non-accountant' (Cambridge Management Training Ltd.) 40 pp.

7.3.3 Management accounting and financial analysis: Books

BROYLES, J., COOPER, I., and ARCHER, S. (1983): 'Financial management handbook' (London, Gower) 456 pp.
CARSBERG, C. (1974): 'Analysis of investment decisions' (London, Haymarket)
FANNING, D. (Ed.) (1983): 'Handbook of management accounting' (London, Gower) 496 pp.
GARBUTT, D. (1985): 'How to budget and control cash' (London, Gower)
SIZER, J. (1985): 'An insight into management accounting' (London, Penguin, rev. ed.) ref. 804412, 520 pp.
SIZER, J. (Ed.) (1980): 'Readings in management accounting' (London, Penguin) ref. 803971, 300 pp.
SIZER, J., and COULTHURST, N. (1984): 'A casebook of British Management Accounting' Vol. I and II (London, Institute of Chartered Accountants, ICAEW)
WILSON, R. M. S. (1982): 'Cost control handbook' (London, Gower, 2nd edn.) 640 pp.
WESTWICK, C. A. (1980): 'How to use management ratios' (London, Gower) 284 pp.
ICC (1985): 'Business ratios' (London, ICC Business Publications)
ICMC (1985): 'Accounting information systems and data processing' (London, Institute of Cost & Management Consultants) ref. 157

7.3.4 Small and growing businesses: Books

HARGREAVES, R. L., and SMITH, R. H. (1984): 'Managing your company's finance' (London, ICFC)
GOLZEN, G. (1984): 'Working for yourself' (London, Kogan Paul/Daily Telegraph)
WOODCOCK, C. (1985): 'The Guardian guide to running a small business' (London, Kogan Paul, 4th edn.) 248 pp.
DREW, R. (1983): 'Microcomputers for financial planning' (London, Gower/ICFC) 116 p.
WILLIAMS, S. (1985): 'The BBC micro and small businesses' (London, BBC)
CA (1983): 'Starting your own business' (London, ConsumerAssociation)
'Reference book for the self-employed and smaller business' (London, Croner Publications) 650 pp. (loose-leaf, with up-date service)

7.3.5 Financial subjects: Sources of training on audio-visual media

JAY, A.: 'The balance sheet barrier'; 'The control of working capital'; 'Cost, profit and breakeven'
HEMINGWAY, J.: 'Budgeting'
WOOLF, E.: 'Depreciation and inflation'
(London, Video Arts) published as video cassettes, with printed notes
Series introduced by WOGAN, T.: 'Valuation of fixed assets and depreciation'; 'Valuation of stocks and work in progress'; 'Inflation accounting' (Maidenhead, McGraw Hill) published as video cassettes
Audio Cassettes on range of business subjects (London, BIM/Audio Magazine Company) issued monthly

EWERT, R. H., WILLIAMS, B. S., and GOERLING, H. K.: 'Fundamentals of finance for non-financial executives' (USA/London, AMR International) series of 10 audio cassettes

Training courses

Management Training Directory (1985/86): (London, Alan Armstrong, 7th edn.) lists over 2000 short courses, run by 170 organisations, on all aspects of management

Courses for the examinations of the Accountancy bodies are available at Polytechnics and Technical or Commercial Colleges, at all principal centres

Courses for the Certified Diploma in Accounting and Finance', intended for managers who are not accountants, are provided by The Chartered Assocation of Certified Accountants (MN), PO Box 244, London WD2A 3EE, tel: 01-242 6855

A range of courses, including a full MBA, are available from The Open University, PO Box 481, Milton Keynes MK7 6BH

7.3.6 Text references: Sections 3.1–3.6

ANDERSON, A. F. (1985): 'William Henley, pioneer instrument maker and cable manufacturer, 1813 to 1882', *IEE Proc.* **132,** Pt. A, pp. 249–261

BAMFORD, W. (1986): 'Industrial electrical instrument makers, 1890–1960', *IEE Proc.* **133,** Pt. A, pp. 125–128

CATTERMOLE, M. J. G. (1987): 'The Cambridge Scientific Instrument Co., 1881–1968', *IEE Proc.* **134,** Pt. A, pp. 351–358

Colloquium on systems engineering – its nature and scope (1986): *IEE Proc.,* **133,** Pt. A, pp. 329–360

Colloquium on systems theory and systems engineering (1988): *IEE Proc.,* 135, Pt. A, pp. 401–418

DRUCKER, P. (1955): 'The practice of management' (London, Heinemann/Pan) Chap 11 470 pp.

RISK, J. M. S. (1982): 'Realism in accounting' *IEE Proc.,* **129,** Pt. A pp. 274–284

FAYOL, H. (1929): 'Industrial and general administration' (Geneva, IMI) (English translation of French text of 1916)

WESTWICK, C. A. (1973): 'How to use management ratios' (London, Gower) 288 pp.

'125 Years of the *Engineer*' (1981): (London, Morgan-Grampian) pp. 159–199

GRAY, R. H. (1983): 'Research and development' *in 'Financial Reporting 1983–*84' (London, ICA), pp. 125–138 (15th year of an annual review)

7.3.7 Text references: Sections 3.7–3.8.2

ANSON, C. (1971): 'Profit from figures: a manager's guide to statistical methods' (London, McGraw Hill) 273 pp.

AMARA, R., and LIPINSKI, A. (1983): ' Business planning for an uncertain future: scenarios and strategies 'Oxford, Pergamon) 228 pp.

BAKER, K. R., and KROPP, D. H. (1985): 'Management science: an introduction to the use of decision models' (New York, Wiley) 650 pp.

BATTERSBY, A. (1966): 'Mathematics in management' (London, Penguin) 256 pp.

BATTY, J. (1975): 'Standard costing' (London, MacDonald & Evans)

BLANNING, R. W. (1980): 'How managers decide to use planning models', *LRP,* **13,** April, pp. 32–35

BRAGG, J. P. (1982): 'Priority base budgeting: a practical zero base approach to manufacturing overheads' *IEE Proc.* **120** Pt. A, pp. 76–80

BUNN, D. (1982): 'Analysis for optimum decisions' (New York, Wiley) 275 pp.

CHANDLER, J. and COCKLE, P. (1985): 'Techniques of scenario planning' (London, McGraw-Hill) 193 pp.

CONWAY, H. G. (1963): 'Design and productivity' *Production Engineer,* May

COYLE, R. G. (1971): 'Mathematics for business decisions' (London, Nelson) 309 pp.

CROSBY, P. B. (1979): 'Quality is free' (New York, McGraw Hill) 309 pp.

HALL, P. (1980), 'Great Planning disasters' (London: Weidenfeld & Nicholson), 308 pp.

FEIGENBAUM, A. V. (1961): 'Total quality control; engineering and management' (New York, McGraw Hill) 627 pp.

GAGE, W. L. (1968): 'Value Analysis' (New York, McGraw Hill)
GENEEN, H. S. (1985): 'Managing' (London, Granada) 225 pp.
JOHNSTON, D. L. (1988): 'Development of the BS5750 concept of total quality system procedures' *Chemistry and Industry,* No. 11, 6 June, pp. 365–368
LEEMHUIS, J. P. (1985): 'Using scenarios to develop strategies', *LRP,* **18,** April, pp. 30–37
LEWIS, D. M. P. (1969): 'A history of value engineering' *Value Engineering,* December, pp. 133–137
MILES, L. D. (1961 & 1971): 'Techniques of value analysis and engineering' (New York, McGraw Hill) 253 pp. and (2nd edn.) 366 pp.
NEUSCHEL, R. F. (1976): 'Management systems for profit and growth' (New York, McGraw Hill) 365 pp.
MOD, (1983): 'Value for money – Defence open government'. Ref. 83/01 (London, MoD Industrial Policy Division)
OUGHTON, F. (1969): 'Value analysis and value engineering' (London, Pitman) 118 pp.
PYHRR, P. A. (1973): 'Zero-base budgeting: a practical management tool for evaluating expenses' (New York, Wiley)
RACAL (spokesman) (1980): 'Management', *IEE Proc.,* **127,** Pt. A, p. 586
RADKE, M. (Ed.) (1972): 'Manual of cost reduction techniques' (New York, McGraw Hill) 288 pp.
SASSONE, P. G., and SCHAFFER, W. A. (1978): 'Cost-benefit analysis' (London, Academic Press) 192 pp.
TURNER, G. (1985): 'GEC's pragmatic planners', *LRP,* **18,** February, pp. 12–18
WALTON, F. L. J. (1987): 'The role of the City', *IEE Proc.* **134,** Pt. A, pp. 825–833
WILLIAMS, L. A. (1985): 'Microcomputers and marketing decisions' (London, IEE/Peter Peregrinus) 273 pp.
BIM Checklist No 53: 'Value Analysis' (Corby, BIM Management Information Centre)

7.3.8 Text references: Sections 3.8.3 and 3.8.4

ALEXANDER, G. S. (1985): 'The English clearing banks' IEE Proc., 132 Pt. A pp. 484–6
BABBEL, D. F. (1982): 'Exchange-rate fluctuations and transaction exposure in the multinational corporation' *IEE Proc.* **129** Pt. A, pp. 261–264
BIENKOWSKI, M., and ALLEN, K. (1985): 'Industrial aid in the UK: 1985' (Glasgow, CSPP, University of Strathclyde)
CHANDLER, A. D., and SALSBURY, S. (1971): 'Pierre S DuPont and the making of the modern corporation' (New York, Harper & Row) 722 pp.
GOLDSMITH, W., and CLUTTERBUCK, D. (1984): 'The winning streak: Britain's top companies reveal their formulas for success' (London, Weidenfeld & Nicolson) 204 pp.
ICSA (1985): 'How to form a company' (Cambridge, ICSA Publishing)
LYONS, N: (1976): 'The Sony vision' (New York, Crown) 235 pp.
NEDO (1986): 'External capital for small firms' (London, NEDO Books)
PETERS, T. J., and WATERMAN, R. H. (1982): 'In search of excellence: lessons from America's best-run companies' (New York, Harper & Row) 360 pp., (and in paperback)
RICH, S. R., and GUMPERT, D. E. (1985): 'How to write a winning business plan', *Harvard Business Review,* **85,** pp. 155–166, May/June
RODGERS, B. (1986): 'The IBM way' (New York, Harper & Row) 235 pp.
ROTHWELL, R., and ZEGVELD, W. (1980): 'Possibilities for innovation in small and medium sized manufacturing firms' *IEE Proc.* **127,** Pt. A, pp. 267–271
SAMPSON, A. (1973): 'The sovereign state of ITT' (London, Hodder & Stoughton) 288 pp.
SCHUMACHER, E. F. (1973): 'Small is beautiful(New York, Harper & Row) 290 pp.
SECKER, P. E. and MACROSSON, W. D. (1981): 'The university-industry interface', *IEE Proc,* **128** Pt. A, pp. 280–290
SLOAN, A. P. (1963): 'My years with General Motors' (London, Sidgwick & Jackson/Pan)
SOBEL, R. (1983): 'ITT: the management of opportunity' (London, Sidgwick & Jackson) 412 pp.
TAYLOR, G. N. (1986): 'Funding of new technologies', *IEE Proc.,* **133,** Pt. A, p. 74

TUGENDHAT, C. (1972): 'The multinationals' (London, Eyre & Spottiswoode) 242 pp.
YOUNG, G. (1985): 'Venture capital in high-tech companies' (London, Pinter)
WESLEY, J. (1985): 'The contribution of the venture capitalist' *IEE Proc.*, **132** Pt. A, pp. 486–7
WINDSOR, E. (1984): 'Subsidies for innovation' *Innovation and Management*, Summer, pp. 2–5

7.4 Legal and social obligations

ARMSTRONG, M. (1984): 'Handbook of personnel management practice' (London, Kogan Page, 2nd edn.) 416 pp.
BAKER, J. W. (1985): 'When the engineering has to stop' *IEE Proc.*, **132** Pt. A, pp. 193–198
BERESFORD-HARTWELL, G. M. (1987): 'Arbitration in mechanical and electrical engineering', *IEE Proc.*, **134**, Pt. A, pp. 343–350
BOYLE, A. J. and BIRDS, J. (1983): 'Company law' (Bristol, Jordan) 824 pp.
BURBRIDGE, R. N. G. (1987): 'Project management' (London: IEE/Peter Peregrinus) 154 pp.
CARSBERG, B. (1986): 'OFTEL: a new approach to regulation' *IEE Poc.*, **133** Pt. A, pp. 152–158
CARSBERG, B. (1986): 'Regulating private monopolies and promoting competition', *LRP*, **19**, December, pp. 75–81
'Croner's reference book for employers' (New Malden, Croner Publications) with update service
CROSBY, P. B. (1979): 'Quality is free' (New York, McGraw Hill) 309 pp.
'Everyman's own lawyer' (1981): (London, Macmillan, 71st edn.; first published 1863)
FLINT, M. F. (1985): 'Auser's guide to copyright' (London, Butterworth, 2nd edn.) 288 pp.
HARVEY, B. W. (1982): 'The law of consumer protection and fair trading' (London, Butterworth, 2nd edn.) 424 pp.
Health & Safety Commission (1985): 'Report 1984–85' (London, HMSO)
HEARN, P. (1986): 'The business of industrial licensing' (Aldershot, Gower) 250 pp.
HEWART, LORD JUSTICE, (1924): 'Rex v Sussex Justices, 9 Nov 1923', *King's Bench Reports*, **1**, p. 259
HIMMELFARB, P. E. (1985): 'Product failures and accidents' (Basel, Technomic publishing) 311 pp.
IoD (1985): 'Guidelines for directors' (London, Inst. of Directors) 96 pp.
IoD (1986): 'Company Insurance' (London, Inst. of Directors) 80 pp.
IEE Professional Briefs (1984): 'Product liability'; Health and safety legislation'; 'Copyright: photocopying IEE publications'
JANNER, G. (1979): 'Janner's Product Liability' (London, Business Books) 406 pp.
JOHN, M. N. (1984): 'Consultancy' *IEE Proc.*, **131** Pt. A, pp. 341–428
MCGOVEN, E. (1982): 'International Trade regulation: GATT, USA and EEC' (Exeter, Globefield Press) 457 pp.
MELVILLE, L. W. (1972): 'Intellectual property and international licensing' (Andover, Sweet & Maxwell Spon, 2nd edn.)
MILLS, G. (1981): 'On the Board' (Aldershot, Gower) 240 pp.
MINOW, N. (1985): Quoted in *Time (US)*, 18 March
MONAT, J., and SARFATI, H. (1986): 'Workers' participation – a voice in decisions 1981–5' (Geneva, London, ILO)
MORSE, G. (Ed.) (1983): 'Company Law – Charlesworth and Cain' (London, Stevens, 12th edn.) 742 pp.
NEALE, A. D. (1970): 'The antitrust laws of the USA' (Cambridge University Press, 2nd edn.) 530 pp.
NIERENBERG, G. I. (1973): 'Fundamentals of negotiating' (Geneva, Management Editions) 306 pp.
REID, B. C. (1984): 'A practical guide to patent law' (Oxford, ESC Publishing) 438 pp.
ROBERTS, D. (1985): 'How to form a company' (London, ICSA) 112 pp.

RUSSELL-CLARK, A. D. and FYSH, M. (Eds.) (1974): 'Copyright in industrial design' (Andover, Sweet & Maxwell Spon, 5th edn.)

SELWYN, N. (1982): 'Law of health and safety at work' (London, Butterworth)

SMITH, K. and KEENAN, D.J. (1979): 'English Law' (London, Pitman, 6th edn.) 818 pp.

WHEELDON, A. J. D. (1988): 'The responsibilities of engineers in performing contracts'; in 'Engineering Manufacturing Systems', London: Peter Peregrinus (to be published).

White Paper (1982): 'Standards, quality and international competitiveness' Cmnd. 8621 (London, HMSO)

7.5 Marketing products and services

Adler, L. (1967): 'Systems approach to Marketing', *Harvard Business Review,* May–June, pp. 105–118

ALBRECHT, K., and ZEMKE, R. (1985): 'Service America' (Homewood, Ill., Dow Jones Irwin) summary in *International Management,* Nov., pp. 65–67

ALLENDER, P. (1984): 'QA works if you do – an introduction to quality assurance', *BSI News,* August, pp. 15–16

AYWYOS (1834): 'Hints on etiquette and the usages of society' (London, Longman; reprinted 1946, Turnstile Press) 68 pp.

BABBEL, D. F. (1982): 'Exchange-rate fluctuations and transactions exposure in the multinational corporation' *IEE Proc.,* **129** Pt. A, pp. 261–264

BAKER, M. J. *et al.* (1983): 'Market development: a comprehensive survey' (Harmondsworth, Penguin) 240 pp.

BARNARD, P. (1984): 'Research in the USA' *J. Market Research Society, pp. 273–29*3

BAYLISS, J. S. (1985): 'Marketing for Engineers' (London, IEE/Peter Peregrinus) 391 pp.

BOOTHE, R. (1985): 'Reducing exposure to exchange rate risk', *LRP,* **18**, June, p. 98

BOTB (1979): 'Industrial market researcher's checklist', (London, BOTB) 21 pp.

BROOKE, M. Z., and BUCKLEY, P. J. (1984): 'Handbook of International Trade' (Brentford, Kluwer) 1000 pp. with update service

BROWN, J. A., and SOMMERS, D. L. (1982): 'Developing a strategic marketing orientation in a large industrial firm', *Industrial Marketing Management,* **11**, pp. 167–171

CALDECOTE, Viscount R. (1986): 'The competitive position of the European telecommunications industry', *IEE Proc.,* **133**, Pt. A, pp. 365–368

CHISNALL, P. M. (1985): 'Strategic industrial marketing' (Englewood Cliffs, NJ, Prentice Hall) 352 pp.

CLEMENT-JONES, J. (1986): 'Intercept strategy', *IEE Proc.,* **133**, Pt. A, pp. 73–74

COLLOQUIUM (1983): 'Design for Safety', *IEE Proc.,* **130**, Pt. A, pp. 292–299

CONSTABLE, G. E. P. (1985): 'BS is proposed for product design', *BSI News,* June, p. 10

COWELL, D. W. (1984): 'Marketing of Services' (London, Heinemann/Inst. Marketing) 340 pp.

DESSAUER, J. H. (1971): 'My years with Xerox (New York, Doubleday) 239 pp.

DREW, J., JUDKINS, P. and WEST, D. (1985): 'Networking in Organisations' (Aldershot: Gower) 152 pp.

DRUCKER, P. (1973): 'Management – tasks, responsibilities, practices' (London, Heinemann) 839 pp.

FOSTER, R. (1986): 'Innovation – the attacker's advantage' (London, MacMillan) 316 pp.

FRANCIS, N. W. (1986): 'System engineering approach to the major capital projects in BP', *IEE Proc.,* **133**, Pt. A, pp. 338–340

GAYTHORPE, R. S. (1984): 'The right tool for the job. Improving export sales', *IEE Proc.,* **131**, Pt. A, pp. 79–84

GENERAL ELECTRIC (1986): 'A venerable giant sharpens its claws', *IEEE Spectrum,* **23**, pp. 54–65, February

GOLZEN, G. (1986): 'Working abroad' (Sheffield, Kogan Page, 9th edn.)

HAKANSSON, H. (Ed.) (1982): 'International marketing and purchasing of industrial goods – an interactive approach' (Chichester, Wiley) 406 pp.

HALLEY, C., BRECH, E., and ROBINSON, H. (1984): 'Marketing for success in the recovery' (Corby, BIM/Management Research Groups) 36 pp.

HARBOTTLE, J. R. (1981): 'Marketing of electrical products' *IEE Proc.* **128** Pt. A, pp. 384–396

HART, N. (Ed.) (1984): 'The marketing of industrial products' (London, McGraw Hill, 2nd edn.) 212 pp.

HARVARD Business Review Library: 'Part IV: Marketing planning and strategy' (London, Heinemann) (reprints of articles from *HBR* by subject: 15 volumes)

HEDLEY, B. (1977): 'Strategy and the business portfolio', *LRP,* **10,** pp. 9–15, February

HITCHINS, D. K. (1986): 'Managing system creation', *IEE Proc.,* **133,** Pt. A, pp. 343–354

HITCHINS, D. K. (1988): 'Systems creativity' *IEE Proc* **135** Pt. A. pp. 407–418

IEE (1982): 'Professional conduct'. Professional Brief (London, IEE) 12 pp.

IEE (1984): 'Product Liability'. Professional Brief (London, IEE) 28 pp.

IEE (1984): 'Consultancy'. Professional Brief (London, IEE) 18 pp.

IEE (1985): 'Working abroad'. Professional Brief (London, IEE) 24 pp.

JOHN, M. N. (Ed.) (1984): 'Consultancy' *IEE Proc.,* **131** Pt. A special issue, pp. 341–428

JONES, A. (1986): 'Taking advantage of technology in the international market place' *IEE Proc.* **133** Pt. A (to be published)

KOTLER, P. (1980): 'Marketing management' (Englewood Cliffs, NJ, Prentice Hall, 4th edn.) 722 pp.

KOTLER, P. (1983): 'Principles of marketing' (Englewood Cliffs, NJ, Prentice Hall, 2nd edn.) 676 pp.

LESKO, M. (1983): 'Information USA' (New York,Penguin) 990 pp.

LEVITT, T. (1960): 'Marketing myopia', *Harvard Business Rev*., July–August, reprinted Sept.–Oct. 1975, pp.

LEVITT, T. (1965): 'Exploiting the product life cycle', *Harvard Business Rev.,* November/December, pp. 81–94

LEVITT, T. (1974): 'Marketing for business growth' (New York, McGraw Hill) 266 pp. Previous edition, (1969) 'The marketing mode'

LOMAS, T. (1983): 'Quality in British Telecom', *B. Telecom J.* Winter 1983, pp. 33–34, and *J. B. Telecom Engrs.,* Spring 1986

McCAFFERTY, D. N. (1980): 'Successful field services management' (New York, American Management Association)

MACLEAN, I. (Ed.) (1976): 'Handbook of industrial marketing and research' (Isleworth, Kluwer-Harrap) 600 pp.

McTAVISH, R., and MAITLAND,A. (1980): 'Industrial marketing' (London, MacMillan) 204 pp.

MAJARO, S. (1982): 'Marketing in perspective' (London, Allen & Unwin) 236 pp.

MAJARO, S. (1982): 'International marketing' (London, Allen & Unwin) 307 pp.

MANN, D. G. (1984): 'Role of the consulting engineer in quality assurance' *IEE Proc.,* **131** Pt. A, pp. 389–396

McLEOD, T. S. (1983): 'The management of computer aids to design, manufacture and testing in a large electronics company' *Computer Aided Engineering J.,* **1**, pp. 32–35

MONDS, F. (1984): 'The business of electronic product development' (London, IEE/Peter Peregrinus) 144 pp.

MORSE, S. (1982): 'Management skills in marketing' (Maidenhead, McGraw Hill) 150 pp.

MORT, D., and SIDDALL, L. (1986): 'Sources of unofficial UK statistics' (Aldershot, Gower) 467 pp.

NELSON, Lord (1984): 'Export competitiveness', *IEE Proc.,* **131,** Pt. A., pp. 626–634

NIERENBERG, G. I. (1973): 'Fundamentals of negotiating' (Geneva, Management Editions) 306 pp.

NORMANN, R. (1984): 'Service Management: strategy and leadership in service business' (Chichester, John Wiley)

OAKES, Francis (1987): 'The Japan Syndrome, Parts 1–3' *Electronics and Power,* Sept/Nov., pp. 580–583, 619–623 & 735–739

PINCHOT, G. (1985): 'Intrapreneuring' (New York, Harper & Row) 368 pp.

PHILIPS GROUP (1986): 'Taking on Japan Inc', *Time*, May 19th, pp. 52–57

REGINALD (*circa* 1170): 'Life of St Godric' (by a monk of Durham)

ROBERTSON, J. (1985): 'Future work' (Aldershot, Gower) 234 pp.
ROBSON, M. (1982): 'Quality circles: a practical guide' (Aldershot, Gower) 216 pp.
RYAN, C. G. (1984): 'The marketing of technology' (London, IEE/Peter Peregrinus) 144 pp. p. 3
SASSOON, D. (Ed.) (1982): 'Bidding for projects financed by international lending agencies' (Aldershot, Gower) 360 pp.
SCOTT, B. (1985): 'The skills of negotiating' (Aldershot, Gower) 256 pp.
SHUTER, R. (1985): 'Assignment America: Foreign managers beware', *International Management,* Sept. pp. 93–95
SPICKERNELL, D. G. (1983): 'Design for quality', *IEE Proc.,* **130** Pt. A, pp. 159–161
STANTON, W. J. (1984): 'Fundamentals of marketing' (New York, McGraw Hill) 697 pp.
TISDALL, P. (1982): 'Agents of change: the development and practice of management consultancy' (London, Heinemann) 163 pp.
TISDALL, P. (1983): 'Marketing for engineers', *Technology,* **19** September, pp. 16–17
TULL, D. S. and HAWKINS, D. I. (1987): 'Marketing research: measurement and method' (London, Collier MacMillan) 794 pp. 4th edition
VERSCHUUR, J. J. (1984): 'Technologies and markets' (London, IEE/Peter Peregrinus) 212 pp.
WELLEMIN, J. H. (1984): 'Handbook of professional service management' (Bromley, Chartwell-Bratt) 225 pp.: also to be published (1987): 'Professional service engineering'
WHEELDON, A. J. (1986): 'Identifying a product strategy', *IEE Proc.* **133**, Pt. A, pp. 623–629
WILSON, A. (1972): 'Marketing professional services' (Maidenhead, McGraw Hill) 193 pp.
WITTREICH, W. J. (1966): 'How to buy/sell professional services' *Harvard Business Rev.* March–April, pp. 127–138
Industrial Marketing Management: Quarterly Journal (Amsterdam, Elsevier Publishing Co.)
Resident abroad (London: *Financial Times*) monthly

7.6 Popular management books

Reviews of recent management books appear at intervals in the *Financial Times*, and in the BIM Newsletter *Management News.*

The latter publishes each autumn a 'top 20' league table of the books most borrowed from BIM's library, which is reproduced here by permission of BIM Information Services, from *Management News* for November 1988. It is a selective guide to the current books that are being talked about, and found to be helpful.

1 PETERS, T. (1988): 'Thriving on chaos' (London: Macmillan). Emphasis on the need for flexibility combined with excellence in management
2 NEDC/BIM, (1987): 'Making of managers'. The Handy Report has stimulated interest in management training
3 LOCK, D. (1977): 'Project Management' 2nd Edition (Aldershot: Gower). Upsurge of demand for a classic text.
4 KHARBANDA, O. P. and STALLWORTHY, E. A. (1982): 'How to learn from project disasters' (Aldershot: Gower). The need to learn; reader-demand prompted by recent disasters: risk-analysis.
5 WOODCOCK, C. (1987): 'The Guardian guide to running a small business' (London: Kogan Page). Sound advice, in its 6th edition since 1980.
6 ROBSON, M. (1986): 'The journey to excellence' (Chichester: John Wiley). Prescribes top management commitment combined with contributions at all levels.
7 PORTER, M. (1980): 'Competitive strategy' (London: Free Press). Techniques for analysis of industries and competitors.

8 HARVEY-JONES, J. (1988): 'Making it happen' (London: Collins). How the author set about galvanising ICI from top to bottom, after entering at a modest level from the Navy.
9 HIRSH, W. and BEVAN, S. (1988): 'What makes a manager' (Falmer, Brighton: I. of Manpower Studies). Identifies the key management skills and attributes as more important than experience.
10 PETERS, T. and WATERMAN, R. (1982): 'In search of excellence' (New York: Harper and Row). A best seller for 8 years, although the 'top' company names change.
11 BARROW, C. and GOLZEN, G. (1987): 'Taking up a franchise' 4th edition (London: Kogan Page). A way to shorten the learning curve of working for yourself.
12 TAPPER, R. (1985): 'Becoming a consultant' (New York: John Wiley). Written for the USA scene, but nevertheless a helpful DIY guide.
13 DYER, W. (1987): 'Team building' 2nd edition (Reading, Mass.: Addison-Wesley). Emphasis on achieving effective interaction and development of a team.
14 SHELLEY, D. and COHEN, D. (1986): 'Testing psychological tests' (London: Croom Helm). Looking at their strengths and weaknesses.
15 Incomes Data Services (1979): 'Guide to shiftwork' (London: IDS). Current upsurge of interest reflects trends to flexible working hours.
16 KATZ, B. (1987): 'How to manage customer service' (Aldershot: Gower). Practical approach to training to satisfy market needs.
17 SACKER, F. and MARTIN, N. (1987): 'Customer care' (London: Industrial Society). 'Level of service' provides competitive edge: how to achieve it.
18 BOYATZI, R. (1982): 'The competent manager' (New York: John Wiley). A model for effectiveness is offered.
19 KENNEDY, G., BENSON, J. and McMILLAN, J. (1987): 'Managing negotiations' 3rd edition (London: Hutchinson). Covers business deals and industrial relations.
20 ELGOOD, C. (1988): 'Handbook of management games' 4th edition (Aldershot: Gower). A guide to participative team training material.

Index

This index serves also as a glossary to what might be termed 'management jargon' and, as mentioned on page 188, a further vocabulary of engineering-related terms in several languages will be found in the Kompas Trade Directories.

'Key words' are useful tools of ideas, as they provide a framework within which concepts can be traced, in a hierarchic structure, such as is familiar in the expansion of database menus.

(a broader introduction to their cultural significance appears in 'Keywords' by Raymond Williams, Fontana 1976, 286 pp).

The names of authors can be found listed alphabetically in the References (Chapter 7), for each of the main sections of this book. Individuals are named in this Index only when they have made a particular contribution.